D0961848

The AMERICAN PEOPLE and SCIENCE POLICY

Pergamon Titles of Related Interest

Barney THE GLOBAL 2000 REPORT TO THE PRESIDENT OF THE U.S.: Entering the 21st Century

Etheredge CAN GOVERNMENTS LEARN?

Golden SCIENCE ADVICE TO THE PRESIDENT

Lundstedt/Colglazier MANAGING INNOVATION: The Social Dimensions of Creativity

Mayor SCIENTIFIC RESEARCH AND SOCIAL GOALS: Towards a New Development Model

Miller CITIZENSHIP IN AN AGE OF SCIENCE: A Public Opinion Survey of Young Adults

Rothschild MACHINA EX DEA: Feminist Perspectives on Technology

Zinberg UNCERTAIN POWER: The Struggle for a National Energy Policy

Related Journals*

BULLETIN OF SCIENCE TECHNOLOGY AND SOCIETY

EVALUATION AND PROGRAM PLANNING

TECHNOLOGY IN SOCIETY

***Free specimen copies available upon request.**

The AMERICAN PEOPLE and SCIENCE POLICY

The role of public attitudes in the policy process

JON D. MILLER

Northern Illinois University

Pergamon Press

New York Oxford Toronto Sydney Paris Frankfurt

Pergamon Press Offices:

U.S.A.	Pergamon Press Inc., Maxwell House, Fairview Park, Elmsford, New York 10523, U.S.A.
U.K.	Pergamon Press Ltd., Headington Hill Hall, Oxford OX3 0BW, England
CANADA	Pergamon Press Canada Ltd., Suite 104, 150 Consumers Road, Willowdale, Ontario M2J 1P9, Canada
AUSTRALIA	Pergamon Press (Aust.) Pty. Ltd., P.O. Box 544, Potts Point, NSW 2011, Australia
FRANCE	Pergamon Press SARL, 24 rue des Ecoles, 75240 Paris, Cedex 05, France
FEDERAL REPUBLIC OF GERMANY	Pergamon Press GmbH, Hammerweg 6, D-6242 Kronberg-Taunus, Federal Republic of Germany

Library of Congress Cataloging in Publication Data

Miller, Jon D., 1941-

The American people and science policy.

Includes bibliographical references and index.
1. Science and state--United States--Citizen participation. I. Title.
Q127.U6M55 1983 306'.45'0973 83-4119
ISBN 0-08-028064-1

Printed in the United States of America

To
Raymond H. Gusteson
and
Mary Elizabeth Schwartz,

teachers and friends

Table of Contents

LIST OF FIGURES ix

LIST OF TABLES xi

FOREWORD xv

ACKNOWLEDGMENTS xvii

Chapter *page*

1. THE POLITICS OF SCIENCE 1

2. SCIENCE POLICY IN THE TWENTIETH CENTURY 4

    Basic Objectives 4
    The Prewar Experience 5
    The War Effort 7
    The Control of Atomic Energy 11
    The Creation of the National Science Foundation 14
    The Growth of Resources 17
    Summary 21

3. A CONCEPTUAL FRAMEWORK 22

    A Stratified Model of Policy Formulation 22
    Political Specialization 25
        Interest Specialization 26
        Issue Specialization 27
        The Structure of Political Participation 28
    Summary 32

4. PUBLIC PARTICIPATION IN SCIENCE POLICY 33

    Science Policy Decision Makers 33
    Science Policy Leaders 34
    The Attentive Public for Science Policy 39
    The Nonattentive Public 47
    General Dispositions toward Science and Technology 49

5. THE SCIENCE POLICY AGENDA 55

The Dynamics of Science Policy 57

6. THE ACQUISITION OF RESOURCES 60

The Adequacy of Support for Scientific Research 60
The Quality of Precollegiate Science Education 65
Industrial Research and Development 67
The International Standing of American Science and Technology ... 68
Summary 70

7. THE PRESERVATION OF INDEPENDENCE 71

The Public Understanding of Science 71
Governmental Regulation of Science and Technology 73
The Creationist Challenge 77
Summary 78

8. PUBLIC PARTICIPATION IN SPECIFIC CONTROVERSIES 79

Specific Issue Attentiveness 79
Specific Issue Policy Views 85
The Resolution of Specific Issue Controversies 90
Personal Participation in Specific Controversies 93
Summary 103

9. THE FUTURE OF PUBLIC PARTICIPATION IN SCIENCE POLICY 105

The Emergence of Attentiveness 106
The Development of Attitudes Toward Organized Science 115
Future Political Participation 122

10. THE FORMULATION OF SCIENCE POLICY IN A DEMOCRATIC SOCIETY 125

Is an Attentive Public Inherently Elitist? 125
Is There an Optimal Size for an Attentive Public? 126
The Utilization and Effectiveness of the Attentive Public 131

Appendix: THE DATA BASE 134

REFERENCES 137

INDEX 141

ABOUT THE AUTHOR 145

List of Figures

Figure *page*

1. Federal Expenditures for Research and Development: 1930–80 18

2. Federal Support for Basic Research: 1950–80 20

3. A Stratified Model of Policy Formulation 23

4. The Development of Attentiveness to Organized Science 112

List of Tables

Table *page*

1. Political Specialization among Adults 29
2. A Profile of Political Specialization Groups: 1981 31
3. Demographic Profile of Leaders of Science and Technology: 1981 35
4. Participation in the Governance of Scientific Institutions: 1981 36
5. Leadership in Professional or Disciplinary Society: 1981 37
6. Membership in National Leadership Organizations: 1981 38
7. Efforts to Influence Public Attitudes on Science Policy: 1981 38
8. Efforts to Influence Other Leaders on Science Policy: 1981 39
9. Interest in Science and Technology Issues 41
10. Perceived Knowledge about Science and Technology Issues 42
11. Identification of the Attentive and Potentially Attentive Publics 42
12. A Demographic Profile of the Attentive Public for Science Policy 44
13. Multivariate Distribution of Attentiveness: 1979 45
14. A Logit Model to Predict Attentiveness to Science Policy: 1979 46
15. Sources of U.S. Influence in the World: 1979 49
16. Public Expectations for Future Scientific Achievements 50
17. General Dispositions toward Science and Technology: 1979 51
18. Public Perception of the Relative Benefits and Risks of Scientific Research 52
19. Multivariate Distribution of Risk-Benefit Attitudes: 1979 52
20. A Logit Model to Predict Risk-Benefit Assessment Attitudes: 1979 53

21. Major Problems for Science and Technology: 1979 56
22. Perceived Influence of the Attentive Public for Science Policy: 1981.. 57
23. Social and Political Behavior of the Attentive Public 58
24. Levels of Political Participation....................................... 59
25. The Relationship between Scientific Research and Economic Growth: 1981 61
26. Percentage Supporting Increased Spending for Selected Programs: 1981 62
27. Willingness to Exempt Science from Budget Cuts: 1981 62
28. Preferred Balance between Basic and Applied Research: 1981 63
29. Funding Priorities within Science and Technology: 1981 64
30. The Quality of High School Science Instruction: 1981 65
31. International Position of U.S. Science and Engineering Education: 1981 66
32. Funding Priority for Science and Engineering Education: 1981........ 66
33. Attitudes toward Industrial Investment in Scientific R & D: 1981 67
34. Attitudes toward Tax Incentives for Industrial R&D: 1981 68
35. Attitudes toward Patent Law Modification: 1981 68
36. International Position of U.S. Science and Technology: 1981 69
37. Perceived Public Distrust of Science and Technology: 1981........... 72
38. Estimated Public Understanding of Selected Scientific Concepts: 1981 72
39. Leadership Assessment of Government Regulation: 1981.............. 74
40. Willingness to Restrain Scientific Inquiry: 1981 75
41. Multivariate Distribution of Willingness to Restrain Scientific Inquiry: 1979 76
42. A Stepwise Logit Model to Predict Willingness to Restrain Scientific Inquiry.. 76
43. Attitudes toward Creationism: 1981 78
44. Attentive Publics for Specific Issues: 1979 81
45. Demographic Profile of Specific Issue Attentive Publics: 1979 82

46. Three Logit Models to Predict Attentiveness to Specific Controversies: 1979 84

47. General and Specific Issue Attentiveness: 1979 84

48. Policy Preference concerning Space Exploration: 1979 86

49. Three Logit Models to Predict Policy Preferences in Specific Controversies: 1979 87

50. Policy Preference concerning Nuclear Power: 1979 88

51. Policy Preference concerning Food Additives: 1979 89

52. Best-Qualified Groups to Resolve Specific Controversies: 1979 92

53. Social and Political Behavior of Specific Issue Attentive Publics: 1979 94

54. Levels of Political Participation for Specific Issue Attentive Publics: 1979 95

55. Participation in a Controversy concerning Space Exploration: 1979 97

56. Three Logit Models to Predict Participation in Specific Controversies: 1979 98

57. Reasons for Not Participating in a Controversy over Space Exploration: 1979 99

58. Participation in a Controversy concerning Nuclear Power: 1979 100

59. Reasons for Not Participating in a Controversy over Nuclear Power: 1979 101

60. Change in Shopping or Eating Habits over Food Additives Issue: 1979 102

61. The Structure of Issue Interest among College Students: 1978 108

62. The Measurement of Attentiveness among Young Adults: 1978 110

63. Demographic Profile of Young Adult Attentive Public for Science Policy: 1978 114

64. A Logit Model to Predict Young Adult Attentiveness to Organized Science: 1978 114

65. Young Adult Dispositions toward Science and Technology: 1978 116

66. A Logit Model to Predict Young Adult Assessment of the Benefits and Harms of Science: 1978 118

67. Young Adult Attitudes on Selected Science Policy Issues: 1978 119

68. Young Adult Preferences for Science and Technology Spending: 1978 121

69. Young Adult Attitudes toward Political Participation: 1978 123

70. Attentiveness to, and Interest in, Science Policy: 1957–1981 127

71. A Logit Model to Predict Attentiveness to Science Policy: 1957, 1979 129

Foreword

Jon Miller's *American People and Science Policy* is an imaginatively crafted and well written study of attitudes toward science and technology issues and policies that will both gratify professional students of these problems, and at the same time be accessible to a larger readership.

The concept of the "Attentive Public" was first suggested in a book that I wrote on public opinion and American foreign policy some thirty years ago. Other scholars have employed the concept and attempted to operationalize it in the intervening decades. But it has had to wait for Jon Miller's studies of the science public for this concept to be brought to earth, and elaborated and validated on the basis of a rich body of evidence. Hence the study contributes to democratic theory, as well as illuminating the politics of science in important ways.

The conclusions of the study are based particularly on four major surveys administered by Jon Miller's team: two of the general public (administered in 1979 and 1981), one of high school and college students (1978), and one of the leadership of the science and technology community (1981). The study establishes the size of the attentive public for science, its growth over time, its social composition, the recruitment and socialization of science attentives in the high school and college years, the areas of agreement and disagreement of the attentive public and the science leadership on policy questions, and finally the significance of these properties of the attentive public for science policy and for democratic politics.

Miller's main argument is that the attentive public for science is a substantial and growing stratum of the population (20% as of 1981), interested and informed about science, which is mobilizable in support of scientific and technological growth. In a sense it is a "reserve army" for the political support of science, since its high prestige in the last decades has made it possible for research and development to get an increasing share of public resources without much political controversy. But given the declining rate of economic growth in recent years, the leadership of science in its struggles for its share of public resources may have increasing need for this motivated, informed, and politically sophisticated mass of supporters.

It has to be a matter of some concern therefore that the attentive public for science is not fully in support of the agenda of the science elite. It would seem to lean toward supporting ''equal time'' for ''creationism'' as well as evolutionary theory in school instruction, and places more weight on applied research than on basic. These are surprising and disquieting findings. It would appear that the scientific establishment has its educational work carved out for it, and Jon Miller's study focuses attention on the nature of the problem and appropriate remedial strategies.

Gabriel Almond
Professor Emeritus
Stanford University

Acknowledgments

The proper acknowledgment of intellectual, financial, staff, and family debts is a difficult task for any author. It is particularly difficult when a book is based—as is this one—on five national surveys conducted over 24 years by three different survey research centers.

First, I must acknowledge major intellectual debts to Gabriel Almond and Kenneth Prewitt. The basic attentiveness model was first proposed by Professor Almond. The extensions and amplifications in this work would not have been possible without his original insights. I am also indebted to Professor Almond for his kind Foreword to this book.

Kenneth Prewitt and I began this project as co-investigators while he was serving as Director of the National Opinion Research Center. Professor Prewitt made major conceptual contributions to two National Science Foundation proposals, to our joint analysis of the 1979 survey, and to the development of the instruments for the 1981 surveys. His election to the presidency of the Social Science Research Council in 1980 brought new administrative and travel demands on his time, precluding his active participation in the authorship of this analysis. I am deeply indebted to Ken for his rich conceptualizations, his extraordinary knowledge of survey research, and his encouragement in this project.

As with all intellectual debts, the lenders bear no responsibility for my subsequent use of their ideas.

Second, I must acknowledge the financial support of the National Science Foundation, without which this project would not have been possible. I am indebted to Andrew Molnar, Ray Hannapel, Mary Budd Rowe, Joseph Lipson, and James Rutherford for their original confidence in and support of (NSF grant SEDR77-18491) the 1978 National Public Affairs Study. I am equally indebted to Donald Buzzelli and Robert Wright of the NSF Science Indicators Unit for their support of the 1979 survey (NSF contract C-SRS78-16839) of adult attitudes toward science and technology and of the 1981 surveys of adult and leadership attitudes (NSF grant 8105662). Beyond necessary financial support, I am most grateful for the advice, counsel, and friendship of these NSF staff officers.

Third, this work was advanced significantly by the support of Northern Illinois University. I am especially indebted to President William R. Monat, whose commitment to the centrality of scholarship in university life has made Northern Illinois University an excellent place to do research. Without his strong and continuing support of our Public Opinion Laboratory, much of the data collection on which this analysis is based could not have been undertaken. I am also indebted to my colleagues in the Graduate School and the Political Science Department for a sabbatical leave for calendar 1982, during which a substantial portion of the actual writing was completed.

Fourth, I wish to acknowledge the advice and constructive criticism of numerous friends and colleagues. James A. Davis, Robert Pearson, Paul Sheatsley, Robert Suchner, and Alan Voelker all provided useful comments on survey questions and/or text at various points in the analytic process. I am most grateful for their assistance.

During the several years of data collection and analysis reported in this book, I have been assisted and supported by a most competent research staff. Thomas Barrington, Janet McConeghy, Janis Gorski, Char Rapoport, Argene McDowell, Cathy Shanks, and Beth Raffety labored far beyond the call of duty to meet numerous external deadlines and to accommodate the ebb and flow of my own work schedule. It is not possible to name all of the shift supervisors, interviewers, coders, and student assistants whose work contributed to the collection of the data for this study, but I am most grateful for their contributions.

Prior to my first book, I can recall reading other authors' expressions of gratitude to their families and remarking on the triteness of those acknowledgments. I was wrong. My wife, son, and daughter have now survived two books and I am increasingly aware and appreciative of their unique and invaluable contributions to my work. I am always in their debt.

Finally, it is important to acknowledge the assistance of over 11,000 randomly selected Americans, who have given generously of their time in the surveys reported in this analysis. Survey research has become so much a part of our modern social lives that we too often take it for granted. Without the voluntary uncompensated willingness of our fellow citizens to share their attitudes with us—and ultimately the nation—modern social science analysis would not be possible. We are all deeply in their debt.

The AMERICAN PEOPLE and SCIENCE POLICY

Chapter 1
The Politics of Science

The twentieth century has witnessed the evolution of the practice of science and technology from a predominantly individual investigator, low-budget, privately financed mode to multidisciplinary, high-budget, publicly financed research teams. Edison's small workshop has been replaced by large corporate research centers like Bell Laboratories and Fermi's pile of lead and sandbags under the west stands of Stagg Field has grown into a network of national laboratories at Oak Ridge, Hanford, Brookhaven, Los Alamos, Argonne, and Batavia. As public awareness of the impact and cost of science and technology has grown, the scope of governmental involvement in science policy decisions has expanded.

Science policy issues now constitute a continuing and important component of the national political agenda. Shen (1975) has estimated that half of the issues considered by the Congress in recent years have had significant scientific or technological components. In recognition of the volume and importance of science policy matters, the House of Representatives has established a permanent Committee on Science and Technology and the Senate has two major subcommittees devoted primarily to science and technology policy matters. In addition, the Congress has established a separate Office of Technology Assessment (OTA) with a professional staff of over 200 to advise on science and technology matters. At the other end of Pennsylvania Avenue, the White House reestablished the Office of Science and Technology Policy (OSTP) during the Ford administration and that office is now buttressed by statutory authority.

While political and social scientists have generally recognized the emergence of science policy issues on the national political agenda, their analysis of public understanding of, and participation in, science policy issues has been largely atheoretical. Numerous scientists and engineers have expressed their conviction that the mass public is incapable of comprehending and evaluating science policy options and raised questions about the appropriateness of normal political processes for the resolution of technical disputes. The proposal for a ''science court'' grew out of this frustration.

Survey research has documented low levels of interest and knowledge about organized science[1] (Davis, 1958; Withey, 1959; Taviss, 1972; Etzioni and Nunn, 1974; Miller, Prewitt, and Pearson, 1980). Only a small proportion of the adult population is aware of the existence of science policy issues and only a portion of those would qualify as scientifically literate. It is the low salience of science policy that sets it apart from the normal course of politics and that requires a different conceptual approach.

The problem is fundamental to the concept of democratic government. The emergence of numerous science policy issues on the national political agenda reflects the growing importance of organized science to the welfare of society. Yet, only a minority of the citizenry has sufficient interest and knowledge to debate and resolve the issues. Inevitably, the issues are defined and resolved outside the electoral process. Policies that will affect all Americans are determined by a relatively narrow segment of the citizenry.

The purpose of this book is to examine and evaluate the structure and efficacy of public participation in the formulation of science policy in the United States. After a brief review of major science policy issues in the twentieth century, the roots of the salience problem will be considered. The salience of science policy matters will be compared to the salience of other important policy domains. A stratified model of public policy formulation that appears to fit science policy will be introduced and explored.

Working within this framework, the general dispositions of the public toward organized science and the specific issues on the science policy agenda will be discussed. The attitudes of science policy leaders and various strata of the public on substantive policy issues and controversies will be examined. The dispositions and attitudes of young adults on these same issues will be studied, and the prospects for the future formulation of science policy discussed. The final chapter will return to the basic problem of formulating science policy within a democratic political system and explore the likely and optimal levels of public participation in the process.

The analysis of public attitudes toward, and participation in, science policy will utilize four primary data sets. First, two national surveys of adult attitudes and participation were conducted in 1979 and 1981 under a contract and a grant, respectively, from the National Science Foundation. These surveys were designed by Miller and Prewitt (1979) to provide an empirical test of the stratified model of public participation that will be utilized in this analysis. A technical description of the sample designs, instruments, and data-collection procedures can be found in the Appendix and in Miller, Prewitt, and Pearson (1980) and Miller (1982b).

[1] The term organized science is intended to refer to both basic and applied science and to technology. After extensive analyses, we are convinced that the great majority of the American people, perhaps 90 percent, do not have a meaningful understanding of the differentiation between science and technology. While this differentiation may be useful for policy purposes, it is not useful in studying public attitudes toward, or participation in, science policy.

Second, the policy views and political participation information concerning science policy leaders are based on a 1981 survey of 287 leaders by Miller and Prewitt (1982). This national survey is the first systematic empirical investigation of the attitudes and political behaviors of the leadership of organized science in the United States. A technical description of this study can be found in the Appendix and in Miller and Prewitt (1982).

Third, the data concerning the science policy attitudes and political dispositions of young adults are derived from a 1978 national survey of high school and college students. The 1978 study was designed by Miller, Suchner, and Voelker and the results of that survey have been reported in *Citizenship in an Age of Science* (1980), which also included a technical description of the sample, instrument, and data-collection procedures.

Finally, the historical comparison data came from a 1957 study of interest in science news that was sponsored by the National Association of Science Writers and conducted by the Survey Research Center at the University of Michigan. Although the data were collected for another purpose, they are sufficiently broad to allow a secondary analysis utilizing the stratified model incorporated in the 1979 and 1981 surveys. For a technical description of the sample, instrument, and data-collection procedures employed in the 1957 study, see Davis (1958).

Since the analysis of these data utilizes some relatively new analytic techniques, an additional word of introduction may be useful. As a matter of approach, most of the empirical arguments in the following chapters will utilize both a traditional percentage distribution presentation and a multivariate log-linear analysis. While the use of both methods may occasionally appear to be repetitious, it is hoped that the analysis will be comprehensible to persons with and without formal statistical training. Since log-linear analysis is a relatively new technique, a brief introductory discussion of the method is provided in Appendix B in Miller, Suchner, and Voelker (1980).

Chapter 2
Science Policy in the Twentieth Century

To understand current attitudes toward science policy, it is necessary to know some of the political history of organized science in the United States. Science policy since the Second World War has been of a fundamentally different character than the preceding years of American history. A brief summary of science policy in the pre-1940 period will provide a backdrop to the events of the war years. The unlikely and lasting effects of the extensive participation of American science in the war program will be reviewed and a series of major postwar science policy decisions will be studied. The decisions selected for examination in this chapter have had a major impact on the structure and substance of science policy in the United States. Undoubtedly some substantively important issues and decisions have been excluded from this review, but the purpose of this chapter is to provide a skeletal history within which to understand the current science policy agenda.

BASIC OBJECTIVES

The basic objectives of the scientific community have been simple and straightforward: independence of inquiry and adequate support to sustain the scientific enterprise. In the best tradition of science, it is a parsimonious agenda. It represents the dream of all interest groups: financial support without direct control.

Of the two objectives, the independence of scientific inquiry from external intervention or control generally has taken priority over all other concerns. It is important to understand that this commitment to the independence of scientific inquiry is directed to intervention by persons or groups from outside the scientific community: governmental or nongovernmental. The commitment of the scientific community to professional and disciplinary peer review, however, remains strong.

The second objective: the search for public funding has grown steadily in importance in recent decades. As the cost of scientific instrumentation and the reliance on large research facilities and teams have increased, numerous areas of science have reached the point where it would be impossible to continue work in that area without the support of these facilities by some institutional source. Current research in basic theoretical physics and astronomy illustrates this dependence.

Keeping these two objectives in mind, it is useful to turn to a review of science policy in the twentieth century.

THE PREWAR EXPERIENCE

In the decades prior to 1940, the major concern of the scientific community was its independence. While history reveals relatively few attempts to significantly interfere with scientific study in the United States, the anti-intellectual, anti-immigrant, anti-Catholic, anti-Semitic attitudes of many of the popular movements of the latter half of the nineteenth century and the first decades of the twentieth century must have given many in the scientific community some occasion for concern. The relatively low profile of organized science during much of this period may have been an effective strategy for the survival of independence. Resource concerns were secondary during this period.

This set of concerns can be better understood by thinking about the nature of basic scientific research during this period. Most basic research was performed in a few universities by a small number of professors, working with a few graduate students. Penick *et al.* (1965) described basic scientific research in the United States in 1930 in these terms:

> The American university had clearly emerged by the 1930's as the home of basic research. . . . Yet the American university was a strikingly recent phenomenon in the nation's experience. It had scarcely begun to take form in 1880, and much of the development of its strong and specialized departments, its laboratories, and its great research libraries came after 1900. The best creative brains of American science found a haven as professors at a small number of universities, where they taught graduate students and performed research supported in part by university funds derived from state or private sources. In part also, university research was supported by the professors themselves, in the sense that they did not render accounting to anyone for their time or for many minor expenditures. They simply did what research their other duties and their own pocketbooks allowed them to do (p. 7).

The Morrill Land-Grant Act of 1862 had provided the incentive for the expansion of state universities and had stimulated the development of expertise in a number of applied scientific fields, illustrated by the growth of agricultural research stations. While this early work was applied in character, it helped build the foundations of basic science departments that would expand quantitatively and qualitatively in the decades ahead.

For virtually all of this period, American science operated in the shadow of European science. Americans seeking an advanced scientific education still went abroad in large numbers prior to 1940. While the productivity of American science had been increasing in the early twentieth century, the dominant position of European science can be seen in the distribution of Nobel prizes. Prior to 1939, American scientists had received only 15 out of 128 Nobel awards in physics, chemistry, physiology, and medicine. In the years after the Second World War, these figures would reverse sharply.

Throughout the nineteenth century and the early twentieth century, there was little contact between the scientific community and the federal government. As suggested above, the anti-intellectual character of much of American politics during those years may have been a signal to the scientific community that less visibility was the wiser course, even at the price of potential support. The only method of federal support at that time was the direct contract for work to be performed, and the image of the government contract was one of substantial red-tape and patronage complications.

The fear of external control was deep rooted and not limited to government. At the turn of the century, the Woods Hole Marine Biological Laboratory was a small and struggling institution. Numerous college professors camped in tents and other temporary structures and put up small amounts of their own funds to support the work of the laboratory. In recognition of the potential value of this institution, the newly formed Carnegie Corporation and another group associated with the University of Chicago offered financial assistance to the Woods Hole scientists, but the group declined the external support because they were fearful of any "conditions" that might be imposed with the funds (Greenberg, 1967).

At the same time, the federal government, especially the Congress, saw little value for the government in the support of science. In the middle of the nineteenth century, it took the Congress ten years to decide to accept the bequest of James Smithson and to create the Smithsonian Institution. Although several government bureaus were involved in applied scientific missions, these grew slowly during the first four decades of the twentieth century. Illustrative of government indifference to science, the American Chemical Society wrote a letter to the secretary of war on the entry of the United States into the First World War offering its services, and the secretary replied "that it was unnecessary as he had looked into the matter and found the War Department already had a chemist" (Conant, 1952).

The Great Depression had a substantial impact on the scientific community, increasing its awareness of the need for resources and stimulating limited and targeted government support. Basic scientific research was housed primarily in a few private universities and the stock market crash destroyed the endowment base of many of those institutions. The California Institute of Technology, for example, had a predepression endowment of $4.2 million, which was earmarked for the development of the physics department. After the crash, the value of the endowment dropped to about $250,000 (Greenberg, 1967). In industry, scientific research was one of the first areas cut as economic pressures mounted. Unemployment among scientists and engineers grew substantially.

In 1933, President Roosevelt appointed a Science Advisory Board (SAB) to recommend a plan to assist the scientific community. Chaired by Karl T. Compton, the president of the Massachusetts Institute of Technology, the panel recommended that the federal government provide $16 million for scientific and technical research over a six-year period. The proposals produced no governmental action and the Science Advisory Board expired in 1935. One of the major reactions of the scientific community to the appointment of the SAB was to view it as a potential interference in the affairs of science. Many scientific leaders agreed that the president should have continued to rely on the National Academy of Sciences, the traditional link between the scientific community and government. Professor E. B. Wilson of Harvard warned that the new presidential advisory committee ''takes away the independence of scientific men or it may take it away and may result in a committee supposedly under scientific auspices being under political necessity of making recommendations politically suitable'' (Greenberg, 1967). Thus, even as funding was becoming a more serious problem, the commitment of the scientific community to its total independence from government remained strong.

Toward the end of the 1930s some federal funds did begin to flow to science, but they were limited in size and focus. By 1938, for example, the Works Progress Administration was supporting 30 projects in physics, chemistry, mathematics, and astronomy. In 1937, the Congress established the National Cancer Institute as a part of the Public Health Service and authorized grants to universities and other medical research facilities for cancer-related research. Although only $220,000 was actually allocated for the 1938-40 period, the program broke new ground by providing for a grant mechanism rather than the traditional government contract.

On the eve of the Second World War, then, American science was lodged primarily in private universities and operated on limited resources from university and endowment income. Land-grant universities were growing in stature, but they were still not major forces in the scientific community. Federal support was increasing slowly but was only a minor contribution to the total effort. The scientific community remained wary of federal control or intervention, and the federal government placed relatively low priority on the support of this small and odd group of reluctant clients.

THE WAR EFFORT

The extensive involvement of American science in the Second World War changed the basic relationship between the scientific community and the federal government and built a foundation of operating procedures that have served as the basis of that relationship for most of the last four decades. To understand this transformation, it is necessary to examine the genesis of scientific involvement in the war effort, the structure of that participation, and the impact of wartime research on the scope and nature of organized science in the United States.

As in the First World War, it was scientists who first approached the government in regard to possible participation in the war effort. With the growing power of Hitler and fascism in Germany and Austria in the 1930s, a number of leading scientists immigrated to universities in the United States. A core group of nuclear physicists—including Fermi, Einstein, Wigner, Szilard, Teller, and von Neumann—arrived in the United States knowledgeable about recent advances in nuclear science and sensitive to the potential military consequences of this new work. In 1939, Fermi and Bohr presented a review of the recent developments in nuclear studies at the Fifth Washington Conference on Theoretical Physics. The impact of this report was substantial in both the academic community and in some government laboratories.

The barriers between science and government did not disappear easily. Having been advised by academic researchers and their own scientific staff about the explosive potential of uranium, the Naval Research Laboratory offered a small contract to the Carnegie Institution of Washington to investigate the power potential of uranium, but the Carnegie Institution decided to use its own resources rather than take government funds.

Frustrated by the inability of the scientific community and the military to collaborate, Szilard and Wigner prevailed upon Einstein to sign a letter to President Roosevelt alerting him to the potential value of uranium for weapons use and asking his assistance in securing an ample supply of uranium for the United States. The Einstein letter has been widely credited for bringing the scientific community into the war effort, but it was only an early stimulus to the discussion. The effective participation of American science in the war program was still months away. The Einstein letter symbolized the still distant relationship between the scientific community and the government. It only asked that the president secure a supply of uranium, that meant access to the Belgian Congo at that time, and to name a personal liaison that could expedite the flow of information between the scientists and the president. The letter did not ask for funding to promote research in nuclear physics, but rather suggested that the liaison person named by the president might help

> to speed up the experimental work, which is at present being carried on within the limits of the budgets of university laboratories, by providing funds, if such funds be required, through his contacts with private persons who are willing to make contributions for this cause, and perhaps also by obtaining the cooperation of industrial laboratories which have the necessary equipment (Greenberg, p. 74).

In response, Roosevelt appointed an Advisory Committee on Uranium, headed by Lyman Briggs, the director of the National Bureau of Standards. The other two committee members were ordnance specialists from the army and from the navy. The committee interviewed a large number of leading physicists and concluded there was some potential but that it was not imminent. A total of $6,000 was provided for the operation of the committee and the support of research pertaining to uranium. Given the bureaucratic barriers to government spending at that time, relatively little of it was actually used.

At the same time, Vannevar Bush, the president of the Carnegie Institution of Washington and a leading scientist himself, recognized the potential of nuclear power and the importance of arranging for wider scientific participation in the war effort. He mobilized James Conant, the president of Harvard; Karl Compton, the president of MIT; and Frank Jewett, the president of both the National Academy of Sciences and Bell Telephone Laboratories. Bush was able to obtain an appointment with President Roosevelt in early June of 1940, just as the German *blitzkrieg* was rolling into France. Bush found the president receptive to the greater involvement of scientists in the war effort. He persuaded Roosevelt to create a structure within which scientists could work with government support, but without any direct control by the military or the government. In late June of 1940, Roosevelt signed an executive order that established the National Defense Research Committee (NDRC), a civilian controlled agency that was to be headed by Bush and was to report directly to the president. A year later, the NDRC and the Committee on Medical Research would be combined into the Office of Scientific Research and Development (OSRD), which would guide the participation of American science in the war.

The structure of the OSRD and its relationship to univeristy and industrial laboratories were major departures from previous government policy. This new relationship maximized the participation of scientists and engineers in the war effort and established important precedents in science-government cooperation that would continue in the decades after the war. In view of the long-term importance of the OSRD to American science policy, it is important to focus briefly on the structural and procedural innovations that marked the new organization.

Structurally, the OSRD represented a commitment to the civilian control of science, even in the midst of a major war. The location of the organization in the presidential circle, with direct access to the chief executive, indicated the importance placed on science in the war effort. This structural decision removed scientific activity from the traditional bureaucratic channels and, thus, allowed more policy latitude than would have been available had OSRD been located deeper in the federal government. This favored position within the federal structure is one that American science has been reluctant to surrender.

The OSRD had few full-time staff and depended almost exclusively on nongovernmental laboratories for the research and development activities it decided to pursue. The OSRD contract opened a new form of cooperation by providing support for the conduct of scientific research to nongovernmental institutions and by allowing those institutions wide latitude in the management of those funds. Both Bush and his deputy Irvin Stewart recognized that it was necessary for the contract to hold the receiving institution accountable for the funds, but the balance of the OSRD contract stressed the independence of the investigator. Stewart (1948) later described the contract in these terms:

> The performance clause [of the contract] was a relatively simple provision. The contractor agreed to conduct studies and experimental investigations in connection with a given problem and to make a final report of his findings and conclusions . . . by a specified date. This clause was deliberately made flexible in order that the contractor would not be hampered in the details of the work he was to perform. The objective was stated in general terms; no attempt was made to dictate the method of handling the problem (p. 19).

This basic format insured vigorous and creative approaches by the scientists and engineers and it became the model for the postwar research grant programs of the federal government.

With the OSRD innovations in place, American science was able to secure the funds necessary for its contribution to the war effort without a major reduction of its independence. From the outset, Bush emphasized that the OSRD was not simply a procurement agency for the military in the war effort but rather would bring to the weapons development process its full creativity. In numerous areas, from radar to the atomic bomb, this creativity produced new weapons systems that were totally unforeseen by the military leadership. The conversion of the military was nearly total, and, in the postwar years, the military would seek to fund and utilize this creativity in the continuing development of weapons systems.

It is important to understand the impact of the wartime research on the practice of science itself. By the end of the war, American science had learned to work in large laboratories and teams, had solved problems in months and years that it had considered for decades, and had created a new set of weapons that changed the nature of warfare. American science had a new self-image and a new self-confidence. Scientists had not found direct cooperation with government to be a threat to their independence.

The scope and magnitude of the change was summarized in the Kilgore Report (1945) on the war experience:

> By 1944, the Government was spending more than $700,000,000 a year on research—10 times more than in 1938.
>
> Much of this Federal expenditure has been through non-Federal facilities. The mobilization of Bell Laboratories is illustrative. In 1939, Bell had contracts for Government research to the extent of $200,000, or about 1 percent of Bell's total research expenditure. By 1941, this had increased to $5,700,000, or about 22 percent

of Bell's total research. By 1943, Bell was working on Government research contracts to the extent of $41,800,000, or 82.5 percent of its total. By 1944, Bell's work for the Government was still 81.5 percent of its total, and Government contracts had risen to $56,000,000.

In the mobilization of science for war, the pattern of Government research expenditures was entirely altered. In 1940, 70 percent of Government research was being done in Government laboratories. Twenty percent was being done in non-Government laboratories. The remaining 10 percent was work of the Department of Agriculture in cooperation with state experiment stations and land-grant colleges.

By the middle of 1944, 70 percent of the greatly increased Government program was being done in non-Government laboratories. Fifty percent of Government research was being done in industrial laboratories and 20 percent in educational institutions and private research foundations.

The placing of Government contracts has intensified the prewar concentration of research in a relatively few laboratories. Preference has necessarily been given to well-established laboratories with the greatest experience and existing facilities. General Motors, for example, was awarded research contracts between 1940 and 1944 totalling nearly $39,000,000. The Massachusetts Institute of Technology received contracts in the same period totalling nearly $56,000,000. California Institute of Technology received more than $40,000,000, Columbia University, $19,000,000, Harvard, $15,000,000.

Of about 200 educational institutions receiving a total of $235,000,000 in research contracts from the Government, 19 universities and institutes accounted for three-fourths of the total. Of nearly 2,000 industrial organizations receiving a total of almost $1,000,000,000 in research contracts from the Government, less than 100 firms accounted for more than half of the total (Penick *et al.*, pp. 99-100).

In summary, American science was able to engage in the war effort without a loss of independence and with substantial new resources unknown in previous years. Unfettered by military or bureaucratic structures, scientists and engineers were able to approach the weapons development challenge with their full creativity, and the results were startling to both the military and to American science itself. By the end of the war, American science had a new image of itself and a new confidence in its ability to solve a wide array of problems.

THE CONTROL OF ATOMIC ENERGY

Atomic energy was the most important symbol of the achievement of the scientific community during the war years. It also raised important new issues about the role and extent of government regulation of science in the postwar years. In a fundamental sense, it posed a basic issue of the extent of scientific independence possible in the context of some of the new science developed during the war. The issue was of major importance and all other science policy matters were largely deferred until this matter could be resolved.

Actually, the postwar control of atomic energy had been raised within the Truman administration by the perceptive Vannevar Bush more than a year before Hiroshima. Bush had recognized the pervasive influence that the new nuclear research would have on both weapons technology and on peacetime science and had initiated a series of discussions with Secretary of War Stimson, who was influential with the president. As early as July 1944, Bush, Stewart, and Conant had developed a two-page outline of a bill for the domestic control of atomic energy after the war (Hewlett and Anderson, 1962). Representative of Bush's perspective on the new partnership in atomic energy, his 1944 memorandum proposed a 12-member commission with broad powers to control: all nuclear materials, the manufacture of nuclear materials, and all nuclear research. Five of the commission members would have been scientists or engineers nominated by the National Academy of Sciences, three would have been civilians (presumably nonscientists) appointed by the president, and the army and navy would have named two members each. Bush and Conant would later decide that they would rather have an all-civilian commission, but this early outline reflects the relatively high degree of collaboration that evolved between the leadership of the scientific community and the military leadership during the war years.

With the use of the two atomic bombs against Japan, the nature of the new nuclear science was public. There was immediate recognition in both the Truman administration and the Congress that it would be critical to address the issue of the control of atomic energy as soon as possible. In the summer of 1945, a bill was drafted by two lawyers from the army that called for a nine-member commission that would have included five civilians and two army officers and two navy officers. On October 3, 1945, President Truman sent a message to the Congress requesting the formation of an Atomic Energy Commission and endorsing the principle of civilian control but leaving the exact composition of the commission to the Congress (Hewlett and Anderson, 1962).

Following the president's message, a bill was introduced by Congressman Andrew J. May and Senator Edwin C. Johnson that would have created a nine-member commission to be appointed by the president with the advice and consent of the Senate. The May-Johnson bill did not earmark positions on the new commission for military officers, but the bill did include provisions that repealed previous restrictions on the appointment of military officers to civilian positions and the employment of military officers by civilian agencies. The bill was introduced as an administration measure and clearly had the support of the War Department. The bill was also supported by several of the leaders of the scientific community, including Bush, Conant, and Robert Oppenheimer.

Opposition to the May-Johnson bill began to grow within the scientific community. A group of scientists who had worked on the Manhattan Project formed a new organization: the Federation of Atomic Scientists. The federation was aware that the May-Johnson bill was essentially drafted by the army and was distrustful of the possibility of a substantial number of military appointments to the commission. They wanted a clearer and stronger mandate for a totally civilian commission. They also wanted the new commission to have less control of basic nuclear research that would be conducted in universities.

In the Senate, a special committee was formed to hold hearings on the issue, chaired by Senator Brien McMahon. After talking with numerous scientific groups and leaders, McMahon drafted his own bill, which incorporated the concept of an all-civilian commission and which limited the powers of the commission to intervene in basic research. The McMahon bill was endorsed by the Federation of Atomic Scientists. Speaking on behalf of the federation, Harrison Davies testified that

> Vigorous research and development in nuclear science must be maintained. The use of the fission phenomenon and its byproducts in physical, chemical, biological, medical, and industrial research, as well as in power development, should greatly enrich our country and indeed all mankind.
>
> [The McMahon bill] will foster such research and development through equitable distribution of fissionable material and of byproduct materials to all research workers, by granting funds to independent research organizations, such as universities, private and industrial laboratories, through the widest possible dissemination of information, and by creating a division of governmental research to insure a complete program. . . .
>
> Independent research is encouraged by the bill. Government research is not preferred above private research in the allocation of funds. We regard this as a highly desirable feature of legislation in the atomic energy field or for that matter in any other scientific field. The full development of science rests upon independent research by many individuals in many laboratories throughout the world. . . .
>
> We wish to go on record most strongly as favoring complete exclusion of the military from any policy-making function on the Commission. By this we do not mean to exclude efficient liaison between the Commission and the armed services. . . . However, it is in the best tradition of American Government that policy be made by civilians (Hewlett and Anderson, pp. 146-47).

As the controversy over the composition of the commission and the role of the military grew, the White House conducted a series of meetings on the subject. After hearing the views of the major groups and individuals, President Truman decided to support the McMahon bill. Adopting the new administration position, Secretary of War Patterson testified before the House committee that the president and the War Department favored the Senate-passed McMahon bill and the House then adopted the bill. The Atomic Energy Commission began business in January 1947.

The controversy over the control of atomic energy set important precedents in science policy. First, the principle of civilian control was retained and the role of the military was limited by statute. Second, the scientific community realized its political power in the postwar environment. While the wartime scientific leaders had endorsed the May-Johnson bill originally, thinking that it was an effective compromise that retained a majority of civilian control but recognized the military mission, the effectiveness of the Federation of Atomic Scientists and other scientific societies in obtaining total civilian control was impressive. It was a level of influence unknown to the scientific community in the prewar years.

THE CREATION OF THE NATIONAL SCIENCE FOUNDATION

The planning for the support of science in the postwar environment began before the timing of the end of the war itself was clear. Not surprisingly, Bush had been thinking about the problem for some time and talking to the leadership of the Roosevelt administration about the need for continuing governmental support after the war. There was unanimous support for the idea that the government should continue its funding of scientific research after the war, but since most of the scientific leadership was busy with pressing war-related work, there was little consensus on the organizational form or focus of this postwar support.

Through the war years, Senator Harley Kilgore and his Senate Subcommittee on War Mobilization had been monitoring the government support of scientific and engineering projects. At the time of the establishment of the OSRD, the administration and Senator Kilgore agreed that the president should proceed with the establishment of the office without prior congressional approval, using his war powers, but that the OSRD would be disbanded at the end of the conflict and that the Congress would have an opportunity at that time to help shape any postwar organization for this purpose. Looking toward that objective, Senator Kilgore followed federal contracting processes closely and held periodic hearings throughout the war.

In 1945, Senator Kilgore's subcommittee issued a report that discussed the impact of the war experience on scientific research in the United States and focused on the need for continuing federal assistance. While the general goal was acceptable to Bush and his colleagues, he was concerned that the Kilgore approach combined basic and applied research into one large organization and that the control of this support would be divided equally between nongovernmental scientists and federal agency heads. He feared for the independence of scientific research and the impact of utilitarian demands on basic investigations.

Using his influence inside the administration, Bush drafted a letter for President Roosevelt that asked the OSRD to undertake a study of postwar science support and to make appropriate recommendations (Greenberg, 1967). Bush convened a panel of distinguished scientific leaders and drafted a set of recommendations. The report, *Science: The Endless Frontier*, was released in 1945 and called for creation of a National Research Foundation.

Bush's case for continuing federal support and for a separate new foundation expressed the core values of the scientific community:

> It has been basic United States policy that Government should foster the opening of new frontiers. It opened the seas to clipper ships and furnished land for pioneers. Although these frontiers have more or less disappeared, the frontier of science remains. It is in keeping with the American tradition—one which has made the United States great—that new frontiers shall be made accessible for development by all American citizens.
>
> Moreover, since health, well-being, and security are proper concerns of Government, scientific progress is, and must be, of vital interest to Government. Without scientific progress the national health would deteriorate; without scientific progress we could not hope for improvement in our standard of living or for an increased number of jobs for our citizens; and without scientific progress we could not have maintained our liberties against tyranny. . . .
>
> [Yet] we have no national policy for science. The Government has only begun to utilize science in the Nation's welfare. There is no body within the Government charged with formulating or executing a national science policy. There are no standing committees of the Congress devoted to this important subject. Science has been in the wings. It should be brought to the center of the stage—for in it lies much of our hope for the future.
>
> There are areas of science in which the public interest is acute but which are likely to be cultivated inadequately if left without more support than will come from private sources (Bush, pp. 11-12).

Reflecting the decades of wariness of political involvement, Bush designed a governance structure that would insulate the new foundation from overt political pressures. The new foundation would have been controlled by a board of nine members appointed by the president. These would be part-time appointments, the members would serve without salary and be reimbursed only for their expenses. The board of the foundation would select a full-time director who would manage the affairs of the foundation.

In Bush's conception, the foundation would have had separate divisions of medical research, natural sciences, national defense, scientific personnel and education, and publications and scientific collaboration. Each of these five divisions would have been managed by a committee of at least five members, who would be appointed by the board. The Bush report suggested that in making these appointments to the division committees, the board should "request and consider recommendations from the National Academy of Sciences" (Penick *et al.*, 1965).

Bush delivered his report to President Truman in July 1945. While the general objectives of the report were well received, several key personnel in the Truman administration had significant reservations concerning the degree of independence implied in the governance of the new foundation. Led by John Steelman, one of Truman's assistants, a new report was drafted that endorsed the postwar support of basic scientific research, but which proposed a much more substantial role for government scientists. Steelman proposed that the new foundation be headed by a presidential appointee and that the foundation create an advisory board composed of nine nongovernmental scientists and nine heads of military and domestic scientific agencies within the government. In Steelman's view, the new federal foundation should be responsible for coordinating the work of the numerous scientific and engineering offices and bureaus in the federal government and for identifying unsolved problems common to these offices and agencies. While the Steelman report clearly called for increased federal support for basic research in universities, it was equally concerned about the quality and coordination of science conducted within federal agencies (Penick *et al.*, 1965).

These two views of the federal sponsorship of scientific research were embodied in two major bills introduced in the Congress. The Kilgore bill followed the basic design outlined in the Kilgore subcommittee report, and its proposed governance structure was closer to the Steelman approach than the Bush recommendations. In contrast, the bill introduced by Senator Warren Magnuson reflected the more independent approach outlined in the Bush report. In addition to the governance issue, the two bills also differed on the prospective patent policy for federally sponsored research. Reflecting its author's long-standing interest in small business, the Kilgore bill provided that patents produced under federal sponsorship would become public property and would be available for use by all businesses, large and small. Bush and his colleagues believed that this policy would discourage applied research and effectively exclude private industry from participation in any postwar program. The Magnuson bill reflected the OSRD procedure of allowing the recipient to hold any patents developed under federal sponsorship but requiring that a nonexclusive royalty-free license be issued to the government for its own use.

The resolution of both the organizational and patent issues, however, had to await the resolution of the dispute over the control of atomic energy. The fight over the formation of the Atomic Energy Commission attracted the efforts of most of the leadership and activists in the scientific community. So involved was the scientific community with the atomic energy issue that the tabling of the research foundation legislation in late 1946 brought little protest from the scientific community.

In 1947, Senator Alexander Smith introduced a compromise bill that incorporated some aspects of both positions, but leaned toward the Bush design in governance. The bill passed the Congress, but President Truman vetoed the legislation on the grounds that it effectively removed control of the foundation from the executive branch (Penick *et al.*, 1965; Greenberg, 1967). In his veto message, President Truman said, in part:

> I take this action with deep regret. On several occasions I have urged the Congress to enact legislation to establish a National Science Foundation. Our national security

and welfare require that we give direct support to basic scientific research and take steps to increase the number of trained scientists. I had hoped earnestly that the Congress would enact a bill to establish a suitable agency to stimulate and correlate the activities of the Government directed toward these ends.

However, this bill contains provisions which represent such a marked departure from sound principles for the administration of public affairs that I cannot give it my approval. It would, in effect, vest the determination of vital national policies, the expenditure of large public funds, and the administration of important governmental functions in a group of individuals who would be be essentially private citizens. The proposed National Science Foundation would be divorced from control by the people to an extent that implies a distinct lack of faith in democratic processes. . . .

The Constitution places upon the President the responsibility for seeing that the laws are faithfully executed. In the administration of this law, however, he would be deprived of effective means for discharging his constitutional responsibility (Penick *et al.*, p. 136).

Three more years would be required for a compromise bill to be developed and passed, but in 1950 a National Science Foundation Act was passed and signed by President Truman. The final bill was largely on Truman's terms, and the director of the National Science Foundation (NSF) became a presidential appointee. The National Science Board (NSB) was also appointed by the president but is limited to making recommendations to the president. In the end, the scientific community obtained a source of on-going federal support for basic research with a large degree of independence, but without the degree of insulation or isolation proposed in the original Bush design. The final bill did follow the Bush position on patents in large part.

By 1950, then, the structure for government support for basic science was in place. In contrast to the wariness of the scientific community toward the government only 10 years earlier, a majority of the active researchers in the scientific community were now ready to accept federal support and to participate in the distribution of that support through an intricate set of peer review committees and councils. The scientific community had achieved, in figurative terms, a ''contract'' that provided for public support for their work without any external controls over the substance of that work.

THE GROWTH OF RESOURCES

Having obtained an acceptable structure within which to work, the major focus of the scientific community for most of the period since 1950 has been the acquisition of resources for science and technology. While it is correct to conclude, in broad strokes, that the scientific community has been successful in this quest for public support, the picture is more complex than has been understood generally by either the scientific community itself or the larger political community. It is necessary to begin with a review of the levels of resources acquired during recent decades.

In considering the level of resources obtained, it is useful to focus on two separate sets of data. First, it is necessary to understand the level of federal support for what is generally called "research and development," which includes basic research, applied research, and development. Given the imprecision of these categories in the first place and the changing accuracy of agency reporting concerning this categorization, the aggregate level of funding is a relatively reliable and interpretable measure. Second, it is useful to examine the support for basic research. Support for basic research was the primary objective of the effort to establish a National Science Foundation in the first place and, as will be seen in subsequent analyses, remains at the heart of the policy concerns of the scientific community. Of the three categorizations included in the aggregate figure, the data for basic research are the cleanest. Most agencies that report funding for basic research recognize what it is and are confident that their legislative authorization and executive mission include it.

Turning to the data on resources, it is possible to provide a general profile of federal support for research and development as an aggregate for the last five decades (see fig. 1). During the decade prior to the Second World War, federal support for research and development was minimal. In *Science: The Endless Frontier*, Bush (1945) reported that federal expenditures for research and development in 1930 totaled only $24 million. Expenditures for research and development reached $41 million by 1937 and $74 million in 1940. With the

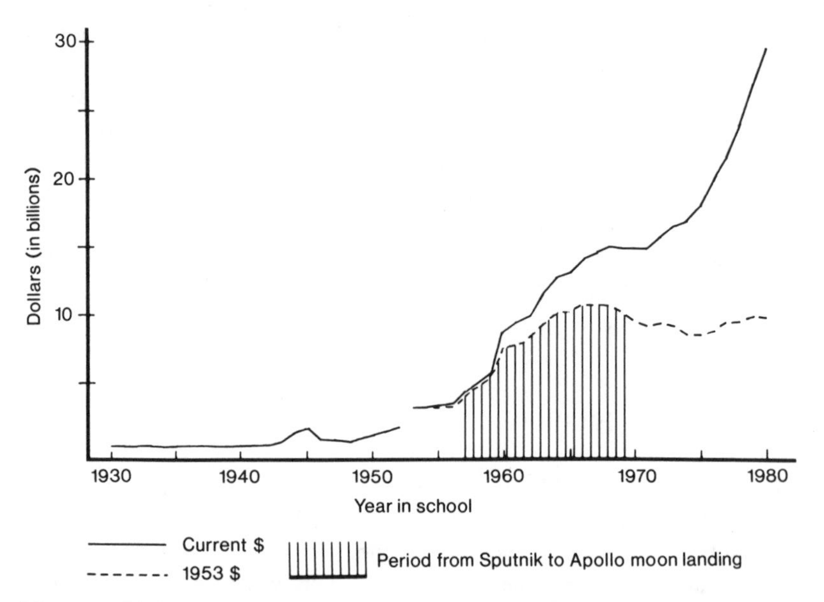

Figure 1: Federal Expenditures for Research and Development: 1930-80

establishment of the OSRD and the later creation of the Manhattan Engineering District for the development of the bomb, the level of federal funds expended for research and development reached $1.5 billion in 1945. The level of funding then dipped below $1 billion until 1950.

The federal definition of "development" was revised in 1953, which created a $1 billion definitional increase in funding levels. This change in definition is shown in figure 1 with a break in the current dollar line. Using the new definition, the level of federal support for research and development was stable from 1953 through 1956, showing an increase from $3.1 billion to $3.4 billion.

On October 4, 1957, the Soviet Union launched Sputnik I and a new era in federal support for research and development. Federal funding for research and development increased by $1 billion in 1957 alone and grew by an additional $10.5 billion (in current dollars) by the time of the landing of the first man on the moon in 1969. It is important to understand that most of this increase in funding was not directly related to the development of space vehicles or even space-related research, but rather the competition engendered by the Sputnik challenge resulted in an increased awareness of scientific activities generally. For example, the Sputnik launch was a sufficient threat to the international dominance of American science that long-term programs like the National Defense Education Act were adopted to finance improved science education in elementary and high schools so as to provide for more competent scientists in the next generation. Often referred to as the golden age of science, this period of about 12 years witnessed a substantial growth rate in federal resources for science.

An analysis of federal support for basic research during the 1950s and 1960s shows a similar pattern of growth. In both statistical and substantive terms, federal support for basic research began after the war. As Bush (1945) pointed out, almost all of the prewar support was applied in nature and the wartime research was largely weapons oriented, although important new physical principles were discovered in some cases. It was not until the new national laboratories created during the war were able to turn their attention to nonweapons problems that the federal support of basic research took on major funding significance. The collection of statistics on the federal support of basic research trailed the event slightly and the most reliable data series, collected by the National Science Foundation, did not begin until 1953. Those data indicate that federal support for basic research totaled $234 million in 1953 and that it had reached $345 million in 1956 on the eve of Sputnik (see fig. 2). By 1969, following the general growth of scientific support and the pro-science environment engendered in large part by the Sputnik challenge, federal support for basic research had risen to $2.4 billion (in current dollars).

These data would suggest, then, that the surge in federal support for research and development generally, and for basic research more specifically, was related in large part to the stimulus of the Sputnik challenge. It should be noted

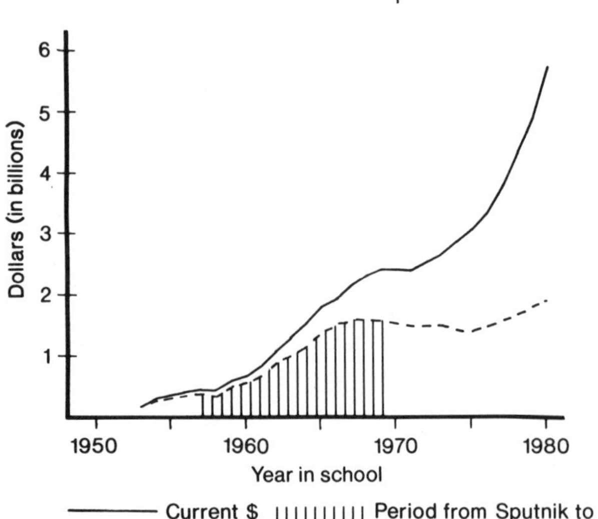

Figure 2: Federal Support for Basic Research: 1950-80

that a significant portion of this growth in resources reflected the expansion of the funding base for the National Institutes of Health (NIH) and it could be argued that this growth in NIH support would have progressed without the Sputnik challenge. It is undoubtedly true that there was growing support for health-related research and that it would have increased at some rate without an external stimulus like the Sputnik launch, but an analysis of the levels of NIH funding over this same period suggests that there was a sharp increase in the rate of growth in the years immediately after Sputnik and that the growth curve for health research funding stabilized in the 1970s in a manner similar to that for basic research. On balance, it seems clear that the launching of Sputnik did play a major role in stimulating a decade of increased federal support for research and development and for basic research.

To this point, all of the federal expenditure figures for research and development and for basic research have been current dollar figures. Given the inflation of the last two decades and especially of the 1970s, it is important to examine these levels of federal support in constant dollar terms as well.

Looking first at the aggregate level of federal support for research and development, the data indicate that when the post-1953 data are displayed in terms of constant 1953 dollars, the rate of growth for the decade after Sputnik is reduced slightly but that all of the increases in the 1970s are eliminated (see fig. 1). In constant 1953 dollars, federal support for research and development increased from $3.1 billion in 1953 to $18.1 billion in 1968 but declined in actual purchasing power in the years following. In 1980, the federal support for research and development was $16.6 billion (in 1953 dollars) or about $1.5 billion lower than the level in the late 1960s.

In the same constant dollars, the federal support for basic research increased from $234 million in 1953 to $2.8 billion in 1968. The level of support then declined for most of the 1970s and reached $2.9 billion again only in 1978 (see fig. 2). In 1953 dollars, the level of federal support for basic research increased to $3.0 billion in 1979 and to $3.1 billion in 1980.

Viewed in constant dollar terms, it is clear that the level of federal support for research and development generally, and for basic research in particular, has been stable for over a decade. The sharply increasing number of current dollars has provided an illusion of growing resources. The reality of stable or declining resources for science and the illusion of substantial budgetary increases have resulted in confusion in the public mind and in the policy process. The period of stable or declining resources is now almost 14 years long and has had a discernible impact on the obsolescence of research equipment, the numbers of graduate students trained, and the vitality of new programs initiated during this period. There has been a visible decline in the morale of the scientific community, that has often concluded in recent years that the public has lost its regard for science or that there is a subtle attempt to renegotiate the basic "contract" between the scientific community and the public. The accuracy and import of these perceptions will be discussed in greater detail in the final chapter. At this point, it is sufficient to note that the cost of this period of stable or declining resources must be examined in broader terms than dollars alone.

SUMMARY

Looking at the major contours of science policy in the twentieth century, it would appear that the experience of the Second World War marked a significant change in the general relationship between the scientific community and the government. The success of the scientific community in creating new weapons for the war effort, especially the atomic bomb, led to new technologies that demanded a stronger and continuing relationship between science and government. In the years immediately after the war, major structural and procedural agreements were developed that provided for the substantive independence of science and for the public support of this work. This basic "contract" reflected a high degree of public and political confidence that science supervised by scientists alone would produce beneficial results for the broader community.

The Sputnik challenge to the preeminence of American science led to a decade of major funding increases for science and technology generally. By the end of the 1960s however, this growth curve began to plateau and the scientific community has experienced more than a decade of stable or declining resources in terms of real purchasing power. Given the growing number of current dollars appropriated for science and technology, the experience of the last decade has created an illusion of increasing support in the face of actual losses. It is within this historical context that current science policy preferences must be viewed.

Chapter 3
A Conceptual Framework

The basic problem in studying public attitudes toward, and participation in, science policy is that organized science is not a salient topic to most Americans. For any citizen, the number of areas of possible interest is vast. One of the characteristics of modern society is that the volume of information is overwhelming and no single individual can become knowledgeable or remain current in more than a relatively narrow range of topics. In marked contrast to our frontier ancestors who waited eagerly for old newspapers from the East, the plight of modern man is to sample selectively from the daily avalanche of information. Inevitably, a process of specialization occurs.

The result of this specialization process is that even those citizens generally interested in political affairs are able to follow only two or three major issues at any given time. Since science and technology policy are relatively complex areas requiring at least some familiarity with scientific terminology, it should not be surprising that most citizens do not choose to follow science and technology matters. Numerous studies (Davis, 1958; Withey, 1959; Oppenheim, 1966; Taviss, 1972; Etzioni and Nunn, 1974) have found a relatively low level of public awareness of science policy issues and a low level of knowledge concerning those issues (Miller, 1983b).

This pattern is not fundamentally different from the levels of public understanding of other important public issues. For example, only a small proportion of the adult population follows foreign policy developments among the nations of the world and even fewer understand the symbolic exchange of words and actions that leads to the formulation of foreign policy in the twentieth century.

The parallel between science policy and foreign policy is both interesting and appropriate for this analysis. Both science policy and foreign policy are complex topics that require a good deal of background knowledge to follow on a day-to-day basis, and both have enormous import to the welfare of the American people.

A STRATIFIED MODEL OF POLICY FORMULATION

Given the apparent contradiction between the expectations of democratic theory and low levels of citizen interest in, and understanding of, foreign policy, Ga-

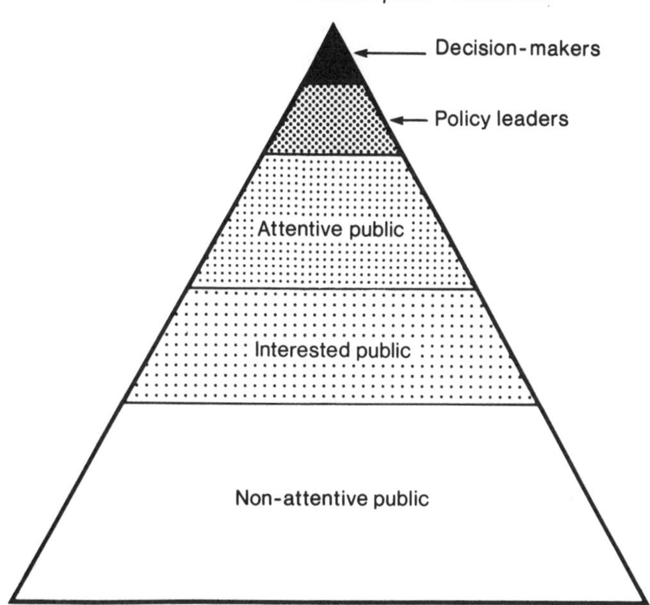

Figure 3: A Stratified Model of Policy Formulation

briel Almond, in 1950, published a basic analysis in his *The American People and Foreign Policy*. Almond argued that the majority of the adult population was generally uninterested in foreign policy matters, becoming interested only in times of war or the activation of the military draft. In more normal times, the public monitoring of foreign policy is conducted by a small portion of the population, that Almond termed an "attentive public" (see fig. 3). There are attentive publics for most low-salience issues.

An attentive public is a self-selected group that has a high level of interest in, and a functional level of knowledge about, a given issue area. Rosenau (1961, 1963) and others (Hero, 1959, 1960; Cohen, 1973; and Mueller, 1973) have provided an empirical description of the attentive public for foreign policy. The concept has been applied to science policy by Miller and his colleagues (Miller, Suchner, and Voelker, 1980; Miller, Prewitt, and Pearson, 1980), and to energy policy and local school policies by Miller (1981, 1983a).

Rosenau (1974) has suggested that an attentive public may be divided conceptually into mobilized and nonmobilized segments. The mobilized public goes beyond interest and information acquisition and seeks to influence policy through overt actions. It is this mobilized group that writes legislators and executives, testifies at hearings, and lobbies decision makers to secure desired policy outcomes. Because an attentive public for most low-saliency policy areas would be relatively small, the mobilized public would be even smaller.

Generally, specific policy objectives are formulated for the attentive public by a still smaller group of policy leaders who are nongovernmental spokesmen for various points of view. For most substantive areas, the policy leadership group includes corporate and labor leaders, academic leaders, and editors of relevant specialized communications media. They debate among themselves and thus help to keep the attentive public informed about the problems and alternatives in any given policy area.

At the top of Almond's public policy pyramid sit the decision makers. The decision makers include those leaders in the executive and legislative branches of the federal government who have the power to make binding decisions. Price (1954, 1965) has provided a perceptive analysis of the structure of science policy decision making in the postwar decades.

Having outlined the structure of the stratified model, it is appropriate to ask how public policy is formulated within this structure. In general, there is frequent policy-related dialogue between decision makers and policy leaders. When there is consensus between the decision makers and the policy leaders on a specific issue, the policy decision is made and there is no wider public participation in the process. In the case of science policy, a large proportion of the matters of greatest relevance to the scientific community are handled through direct negotiations between the policy leaders and the decision makers. The results of these negotiations may be presented at annual meetings and through society or association newsletters, but there is little, if any, effort made to involve the attentive public in the policy negotiation process.

When there are disagreements among the members of the policy leadership group, one or more segments of the leadership group may turn to direct appeals to the attentive public for support. The support requested most often is personal and knowledgeable letters, telephone calls, and visits to decision makers. Occasionally there may be requests for funds to help broaden the appeal and reach other attentives, but this has been somewhat less frequent in the science policy area than in other policy areas. The process of persuading members of the attentive public to make contact with a decision maker to press for a specific policy outcome is referred to as mobilization. At a later point in this analysis, the rates of mobilization of the attentive public for science policy will be estimated for selected issues and disputes.

When there is consensus among the policy leadership group but a reluctance on the part of key decision makers to agree to the demands of the policy leaders, policy leaders may be expected to seek to involve the attentive public and to bring direct pressure to bear on the reluctant decision makers. This type of mobilization has been relatively infrequent during most of the postwar decades.

In a few instances, a segment of the policy leadership group that has been unsuccessful in obtaining adequate support within that stratum or from the attentive public has turned to the general, or nonattentive, public. The best example of this process is the series of popular referenda on nuclear power that have appeared on the general election ballots of various states over the last decade. This type of appeal to audiences considered to be largely uninformed by most members of the scientific community has not been viewed positively by the policy leadership group or the decision makers.

In summary, the stratified model provides a useful framework for understanding the formulation of public policy in low salience areas. Almond argued convincingly that this pyramid of specialization from an uninterested general public, to an attentive public, to nongovernmental policy leaders, and to decision makers does not mitigate the basic tenets of a democratic society but rather reflects the differential ability of individuals in our society to devote the necessary resources, especially time, to become and remain informed and active in numerous policy areas. The total public always retains a final veto power.

POLITICAL SPECIALIZATION

The stratified model needs to be viewed in the context of the broader process of political specialization. The need for specialization springs from a combination of two basic forces. First, citizens in the latter part of the twentieth century are faced with a large number of competing demands for their time, that remains essentially a zero-sum situation. There is ample evidence of the growth of time pressures over the last several decades and no evidence of a reversal of this trend. Faced with a large number of competing and attractive demands on time, fewer adults chose to devote time to political affairs, resulting in a steady decline in public participation in the political system over the last four decades. Even presidential elections, which command the highest levels of public concern and participation, attract barely half of the eligible adults in the United States. This process of allocating time among competing demands will be referred to as interest specialization.

Second, during the same period of time, the information threshold for many political issues has been increasing. Substantial specialized information is necessary for knowledgeable participation in a wide range of political issues. Science policy matters fall into this category, as do most of the issues on the national political agenda. Even an area like economic policy, the traditional pocketbook issue, has become increasingly complex and is beyond the command of a substantial majority of American adults. If only a third of the adult population claim to have a clear understanding of a term like GNP (Gross National Product) (Miller, Prewitt, and Pearson, 1980), how many citizens might be expected to comprehend the current debate over "supply-side economics" or the fate of the dollar in international monetary markets? The result of growing time pressures and rising information thresholds has been the narrowing of the political horizon of most American adults to only a small number of political issues. The process of selecting the issues which an individual wishes to follow will be referred to as issue specialization.

Interest specialization and issue specialization operate simultaneously and their joint effect will be referred to as political specialization. Since subsequent analyses of public attitudes toward, and participation in, science policy formulation will be premised in large part on the political-specialization process, it is necessary to turn first to an examination of the interest-specialization and issue specialization processes and their combined effects.

Interest Specialization

Every citizen must choose whether or not to allocate time and other resources to become and remain informed about political affairs and to participate in the political system. This is not an easy decision and it has been taken for granted, too often, in analyses of citizenship behavior.

For the individual, time is a zero-sum situation and a given hour or day can be allocated only once. In earlier decades when society was predominantly rural and the pace of events was slower, it was possible for literate people to read a substantial portion of the available political news and to find time to discuss political issues of interest. Modern society, however, generates substantially greater and more diverse demands on the individual's time. Most people now live in urbanized settings and have the choice of a wide array of entertainment, numerous clubs and organizations, volumes of printed material on a wide range of topics, sporting events for both personal and electronic viewing, and opportunities for personal involvement and expression in a wide range of creative arts and handicrafts. Bronfenbrenner (1970) has noted the absence of the extended family and the transfer of child rearing from parents to institutions and professionals outside the home. Keniston (1977) has described the effect of these changes on family life and Toffler (1970) has popularized the stress-producing character of the wide range of personal choices available in modern life.

Facing this array of opportunities, the individual citizen must decide between allocating time to political affairs or to other competing attractions. The record of public participation in both electoral and nonelectoral activities suggests that politics has not been a major interest for some time. This does not mean that most citizens are willing to renounce on a permanent basis their right to vote or to make their views known if they should become concerned about an issue or problem; but for the year-in year-out cycles of political events, most citizens are simply not as attracted to political activities as they are to other activities.

Conceptualized in this manner, the general observation is verified in numerous data sets and in a wide array of social science literature. Milbrath (1965) observed that:

> About one-third of the American adult population can be characterized as politically apathetic or passive; in most cases, they are unaware, literally, of the political part of the world around them. Another 60 percent play largely spectator roles in the political process; they watch, they cheer, they vote, but they do not battle (p. 21).

Campbell *et al.* (1960), Converse (1964, 1970), and Verba and Nie (1972) have documented the low level of interest in politics and the even lower level of substantive political information held by a significant segment of the adult population.

A trichotomous measure of the salience of politics was constructed, utilizing the previous conceptualizations of Milbrath (1965) and Verba and Nie (1972). The adult population was classified into three groups:

1. Persons who reported no political activity of any kind
2. Persons who reported voting in a national and/or local election, but reported no other political activity
3. Persons who reported one or more of the following activities: working for a political party or candidate, attending a political meeting, asking someone else to vote for a party or candidate, wearing a campaign button, displaying a political poster, or contacting a public official on a policy issue.

The 1979 data indicated that 23 percent of the adult population were politically inactive, 28 percent were voters only, and the remaining 50 percent found politics to be sufficiently salient to engage in one or more political activities other than voting. This result is consistent with previous studies of political participation and with the interest-specialization process outlined above.

Issue Specialization

A second important decision faced by each citizen is which areas of public policy, if any, are sufficiently important to follow. Given the range of active political issues at the federal, state, and local levels, no individual has the resources or ability to become and remain adequately informed about a very wide range of policy domains or issue areas. In most western industrial nations, it is now imperative that the active citizen limit his or her attention to a smaller number of issues or areas.

Political and social scientists have been slow to understand this imperative. In a classic analysis of mass belief systems, Converse (1964) attempted to find consistency among individuals' attitudes on federal employment policy, aid to education, federal housing policy, the role of the Fair Employment Practices Commission, economic aid to underdeveloped nations, military aid to other nations, and the appropriate degree of United States involvement in world affairs. Converse found low levels of consistency among the attitudes of the public on these issues and concluded that the public lacked the broader and more abstract ideological frameworks necessary for organizing a diverse set of policy views. The issue specialization model offers a different explanation. It would be unlikely to find a very large segment of the population with cogent attitudes on seven diverse policy areas. The absence of consistency might reflect a narrower range of issue attentiveness. So interpreted, the Converse data are consistent with the issue specialization model.

Utilizing data from national surveys conducted in 1979 and 1981[2], it is possible to examine the distribution of attentiveness in several policy domains. Respondents were asked to indicate their level of interest in a set of nine issues[3] and to estimate their own knowledgeability about those issue areas. Following the concept of attentiveness described above, an individual had to declare himself to be "very interested" and "very well informed" about an issue to be classified as attentive to that issue. The subjective nature of the information judgment is particularly important. It is the individual's perception of the adequacy of his information that will affect his disposition toward participation, not an objective external measure. Using this method, the 1979 data indicated that approximately half of the adult population were not attentive to any of the public policy issues included in the study, 23 percent were attentive to a single issue, and 27 percent were attentive to two or more issues.

The Structure of Political Participation

Political specialization refers to the joint effects of interest and issue specialization and an understanding of these effects is a prerequisite for the subsequent analyses of public participation in the formulation of science policy. Using three levels of interest specialization and three levels of issue specialization, a typology of political behavior was produced (see table 1). The three levels of the salience of politics reflect the effect of interest specialization. The number of issues to which an individual is attentive reflect the effect of issue specialization.

The lowest level of activity in the political system was represented by those individuals for whom politics was not salient. This group included one in five adults (see table 1). As implied by the political-specialization construct, most

[2] The technical characteristics of the three surveys utilized in the data analyses reported in this book are described in the Appendix.

[3] In the 1979 and 1981 surveys, respondents were asked to report their level of interest in international and foreign policy issues, agricultural and farming issues, local school issues, issues about new scientific discoveries, economic issues and business conditions, issues about the use of new inventions and technologies, women's rights issues, and issues about energy policy. In 1979, the battery included minority rights issues, which was replaced with space exploration in the 1981 survey.

TABLE 1
POLITICAL SPECIALIZATION AMONG ADULTS

	Number of Issues Followed		
Salience of Politics	0	1	2-7
Not salient	Apolitical Observers 1979: 15% 1981: 9%	Single-issue Observers 1979: 4% 1981: 3%	Multi-issue Observers 1979: 4% 1981: 2%
Vote only	Ritual Voters 1979: 17% 1981: 12%	Single-issue Voters 1979: 6% 1981: 6%	Multi-issue Voters 1979: 5% 1981: 5%
Salient	Ritual Activists 1979: 19% 1981: 17%	Single-issue Activists 1979: 13% 1981: 16%	Multi-issue Activists 1979: 18% 1981: 30%

N (1979) = 1635
N (1981) = 1540

of the persons who did not find politics salient were not attentive to any of the issue areas. A few of these "observers," however, were attentive to one or more of the issue areas. In subsequent analyses, all three of these cells will be treated as one group and referred to as "inactive citizens."

At a second level of participation, about a quarter of the adult population reported voting in local, state, or national elections but indicated no other political activity and no contacting of decision makers. According to the 1979 and 1981 data, many of the vote-only group were not attentive to any of the issues, although a significant segment was attentive to one or more issue areas[4]. In subsequent analyses, these single- and multi-issue voters will be treated as one group and referred to as "issue voters." Over half of the population reported some type of political activity beyond voting: wearing a political button, attending a rally, persuading others to vote for a specific candidate, contributing funds to a political campaign, or contacting decision makers on a policy-related matter. One segment of this active group, representing about 18 percent of the adult population, was attentive to none of the policy areas included in the 1979

[4] In the context of recent debates about issue voting (RePass, 1971; Nie and Anderson, 1978; Sullivan, Piereson, and Marcus, 1978; Nie, Verba, and Petrocik, 1976), it is interesting to note that the number of ritual voters was slightly larger in both years than the number of issue voters, although the difference appears to be diminishing.

and 1981 studies. A second group of activists was attentive to a single issue. Those respondents for whom politics were salient and who were attentive to two or more policy areas represented the classic conception of democratic citizenship. They were active in the political processes of their society and they managed to stay informed about some set of public policy issues.

The grouping of the inactives and the issue voters as described above produced a six-category typology of political behavior that will be useful in understanding the role of an attentive public in the formulation of public policy. The preceding discussion has been descriptive by design, but the survey data employed to describe the basic distribution raise two interesting analytic questions that require a brief examination.

First, while there was a strong positive association between the salience of politics and the extent of issue attentiveness in both years, the fit was less than perfect. Some respondents reported no interest in politics but attentiveness to one or more issues. Other respondents reported politics to be personally salient but were not attentive to any of the issues in the two studies. What might explain these apparently deviant patterns?

Second, the distribution of the population changed significantly for some categories of the typology between 1979 and 1981. For example, the percentage of inactive citizens dropped from 23 percent to 14 percent and the number of multi-issue activists increased from 18 percent to 30 percent over the two-year period. What might explain these changes?

Recognizing that both of these questions might be better answered with longitudinal data, which are not available, it is possible to use the 1979 and 1981 studies to advance some tentative answers. Looking first at the fit of the typology, there were two problem areas: inactive citizens with an interest in one or more policy areas and politically active citizens without a substantive policy interest. An examination of the individual cases involved in these two groups indicated two explanations. Those citizens who reported a high level of substantive interest in a policy area but no interest in political affairs tended to be younger adults who were either unconvinced of the efficacy of political activity or too busy with personal, occupational, or educational events to allocate time to political activities. While the interview schedule did not ask about expected future political activity, the interview responses suggest that most of the young people in this group will become politically active later in the life cycle. In contrast, about half of the ritual activists were 45 years of age or older, and 72 percent of this group were regular or strong political partisans (see table 2). These older adults have been active in partisan politics for several years, perhaps initially motivated by concern over a particular issue, but have continued their political involvement without reference to specific policy matters. Such individuals typify partisanship as a social group activity rather than as an instrument to mold policy.

While these profiles of the populations of the deviant categories have been

TABLE 2
A PROFILE OF POLITICAL SPECIALIZATION GROUPS: 1981

	Inactive	Ritual Voters	Issue Voters	Ritual Active	Single-issue Active	Multi-issue Active
Age						
17–24	33%	10%	13%	14%	16%	16%
25–34	21	22	26	19	26	24
35–44	12	15	16	18	24	17
45–54	10	18	18	20	13	18
55–64	10	12	13	9	5	13
65 and over	14	23	14	20	16	12
Education						
Less than high school	38	29	18	18	16	13
High school	55	65	65	72	58	51
Baccalaureate	4	6	13	9	20	24
Graduate degree	3	0	5	1	7	13
Partisanship						
Strong	13	28	20	31	33	35
Regular	31	45	36	41	31	30
Leaners	14	5	16	10	19	16
Independents	43	21	28	17	18	20
N (1981) =	218	184	174	267	238	459

primarily descriptive in nature, they indicate that the exceptions are comprehensible within a more general understanding of the political process. It is appropriate to conclude that these deviations do not represent a major flaw in the basic typology of political behavior under conditions of political specialization.

The problem of significant differences in the distribution of the population in the typology over only a two-year period also raises questions. The answer appears to rest, in part, in the period covered by the two surveys and, in part, in the dynamic character of political behavior itself. The 1979 study was conducted at the end of the third year of the Carter administration and almost nine months prior to the nominating conventions for the 1980 presidential election. Both houses of the Congress and the executive branch were controlled by the same party, and the incumbent president had indicated his expectation to run for reelection. It was not a period of high political drama. The 1981 survey was conducted toward the end of the first year of the Reagan administration. The control of the Congress was now divided and the new administration had proposed a series of program and budget cuts focused on domestic programs. There was open debate about the future of the Social Security system and the deadline for the ratification of the Equal Rights Amendment was approaching. In relative terms, it was a period of high political drama.

It would have been more surprising to have found no significant changes in the distribution of the population given the two periods in which the data were collected. While changes of 2 to 3 percent can be accounted for in part by sampling error, the larger changes reflect the dynamic character of political behavior under conditions of political specialization. Although some portion of the adult population chooses to follow interests other than politics, it is essential to recognize that these individuals are not permanently lost to the political system but, rather, can reenter whenever they perceive their interests to be threatened or when they perceive that their support might be important to a valued policy outcome. These data from the 1979 and 1981 studies indicate that the levels of attentiveness and participation can adjust rapidly to changed political circumstances. In general, this result does not challenge the basic typology but rather demonstrates the dynamic character of the process.

SUMMARY

In summary, the stratified model of policy formulation first introduced by Almond appears to provide a useful framework for the analysis of public participation in low-salience issues. When viewed in the context of the broader political-specialization process, the roles of an attentive public and of nongovernmental policy leaders become clearer. The 1979 and 1981 data appeared to fit the political-specialization model and to provide empirical descriptions of the interest and issue specialization processes.

The political-specialization model produced a useful six-tier categorization of political behavior. The data from the two surveys indicated that about 30 percent of the adult population of the United States fit the classic model of a democratic citizen, who displays interest in, and knowledge about, several political issues and who takes a more active political role than voting alone. Another 30 percent of the adult population limited their participation to voting, but only a third of this voting group reported any substantive issue interests, suggesting that the remainder may be ritual voters who perform the voting act without an interest in, or understanding of, any substantive issues. In general, the data from both the 1979 and 1981 surveys point to a relatively high level of ritual participation.

Having provided a general framework for the examination of public participation in the formulation of public policy, it is appropriate to turn to the process of formulating science policy and to examine the membership and functions of the strata in the system.

Chapter 4
Public Participation in Science Policy

The preceding chapter outlined a stratified model of public participation in low-salience issues, set that model in the context of a broader political-specialization process, and suggested conditions under which various types of public participation in the policy process would tend to occur. It is now appropriate to turn to an examination of the membership of each of the strata in the stratified model as applied to the formulation of science policy.

SCIENCE POLICY DECISION MAKERS

In the stratified model, the decision makers are those persons or groups of persons who have the power to make binding decisions on policy matters in any given area. The principal decision makers relevant to the formulation of science policy include the leadership of both houses of the Congress and the president and the major officers of the executive branch concerned primarily with matters relating to science policy. In rare instances, federal judges and legislative, executive, and judicial officers at the state and local levels can become decision makers. It is a small group, fewer than 100 persons, but one that has grown significantly since the end of the Second World War.

In the American system of government, the separation of powers means that there is not a single hierarchical set of decision makers. It is possible, if not likely, that there will be differences in the policy preferences of the relevant science policy decision makers on any given issue. These differences usually are resolved through normal negotiation processes, often with the active assistance of key nongovernmental policy leaders.

Most of the decision makers have responsibilities broader than science policy. One of the tasks of nongovernmental policy leaders is to get decision makers at the higher levels to focus on science policy issues. The discussion in recent years over the role of the science advisor to the president may be viewed as a dispute over the level of access that the scientific community should have to the president. The establishment of the Office of Science and Technology (OST) by President Kennedy was a reflection of his desire to have direct and regular communication with the scientific community, and the selection of Jerome Weisner to head the OST symbolized an open channel of communication. In contrast, the abolition of the OST by President Nixon reflected his desire for a more formal and distant relationship with the scientific community. To reestablish the office and prevent its abolition in the future, the Congress gave statutory status to an Office of Science and Technology Policy (OSTP) in 1979. By selecting George Keyworth, a scientist who had not been a principal in the science policy leadership group, President Reagan has attempted to utilize the position as a communication filter.

The Congress, on the other hand, has generally recognized the growing volume and importance of science policy matters and adapted its structure accordingly. The House of Representatives has a permanent Committee on Science and Technology with a substantial subcommittee structure and adequate staffing. The Senate has continued to rely primarily on two major subcommittees but has provided specialized technical staff to assist in the review of science policy matters. In addition, the Congress has established an Office of Technology Assessment (OTA) that studies science and technology policy issues as assigned by a congressional oversight committee. With a professional staff of over 200, the OTA provides a counterbalance to the technical expertise available within executive branch agencies and gives the Congress an independent assessment of executive branch proposals.

In summary, the decision makers for science policy are a relatively small group of legislative and executive officers who can, individually in some cases, but usually collectively, make binding decisions concerning science policy matters. In some cases, the decision makers may include federal judicial officers or legislative, executive, or judicial officers from state and local governments; but the instances of decision making at those levels are not frequent. In general, the Congress has recognized the importance of science policy and provided committee and staff structures to facilitate the decision-making process. The place of science policy in the executive branch has varied from administration to administration.

SCIENCE POLICY LEADERS

Nongovernment science policy leaders play an essential role in the formulation of public policy relevant to science and technology. In contrast to the decision makers, most science policy leaders focus almost exclusively on science and technology matters. There are about 3,000 individuals in the nongovernmental leadership group for science policy. To a great extent, the science policy agenda is formulated by this leadership group. The presentation of policy objectives to the decision makers and any subsequent negotiations are conducted by the policy leaders. As noted above, when there is consensus among the policy leaders themselves and between them and decision makers, policy is established without wider public involvement.

Who are these nongovernmental leaders of science policy? What are their backgrounds? How do they exercise this leadership role?

A 1981 survey of science policy leaders[5] indicated that the leadership group was broadly representative of the major disciplines and professions within the scientific community, predominantly from universities but with significant representation from both the for-profit sector and from independent research centers, and predominantly male with a modal age in the early 50s (see table 3). The distribution of disciplines and professions across sectors was relatively stable, with a slightly higher proportion of engineers and other professionals and a slightly lower number of social scientists in the for-profit sector. The predominance of males in the group was most pronounced in the for-profit sector. The distribution of leaders by age and region of residence was relatively uniform across all sectors.

The policy leaders have a strong organizational base within the scientific

TABLE 3
DEMOGRAPHIC PROFILE OF LEADERS OF SCIENCE AND TECHNOLOGY: 1981

	Total	For Profit	University	Other Nonprofit
Age				
Under 45	27%	21%	30%	23%
45–54	33	40	39	38
55–64	27	31	22	30
65 and over	14	10	19	9
Gender				
Female	12	3	14	13
Male	88	97	86	87
Discipline				
Biological Science	15	13	17	15
Physical Science	25	21	27	26
Social Science	20	9	26	15
Engineering-Prof	20	30	16	23
Other	20	27	13	21
Region				
Northeast	28	32	32	16
North Central	19	22	21	10
South	29	25	20	54
West	24	20	27	19
N =	281	59	147	67
% =	100	21	52	24

[5] A complete description of the technical characteristics of the 1981 survey of science policy leaders is provided in the Appendix.

community. Almost half were members of an editorial board of a scientific or engineering journal (see table 4). Membership on the editorial boards of journals was most frequent among leaders from universities and relatively rare among leaders from the for-profit sector. Approximately a quarter were members of a governing board of an independent laboratory or research center, and about the same number were members of a board of directors of a for-profit corporation involved primarily in scientific or engineering activities. Sixteen percent were members of a governing board of a private foundation engaged in supporting scientific research or development, while only 11 percent were members of a board of trustees of a college or university.

The 1981 data indicated extensive leader participation in national professional and disciplinary societies, with 42 percent reporting that they were, or had been, a national officer of a scientific or disciplinary society. This experience was significantly less frequent among leaders from the for-profit sector and significantly more frequent among leaders with backgrounds in the physical sciences (see table 5). Slightly over half of the leaders had chaired national committees of their professional or disciplinary societies, and almost 70 percent

TABLE 4
PARTICIPATION IN THE GOVERNANCE OF SCIENTIFIC INSTITUTIONS: 1981

	Percentage Who Are a Member of . . .					
	A	B	C	D	E	N
Total	48%	27%	27%	16%	11%	281
Sector						
For-profit	17	24	41	17	20	59
University	60	27	23	14	8	146
Nonprofit	46	30	24	19	12	67
Discipline						
Biological Sci.	64	19	19	12	5	42
Physical Sci.	57	31	22	15	77	69
Social Sci.	50	32	24	9	9	54
Engineering-Prof	36	23	43	25	21	56
Other	37	24	26	17	13	54
Region						
Northeast	49	38	26	19	14	77
North Central	50	18	26	16	14	51
South	39	27	28	19	13	80
West	57	20	26	9	5	65

A = an editorial board of a scientific or technological journal
B = the governing board of any independent center or laboratory that engages in scientific research or development
C = a board of directors of any corporation involved in science or technology
D = the board of directors of any foundation that provides funds for scientific research and development
E = a board of trustees of a college or university

TABLE 5
LEADERSHIP IN PROFESSIONAL OR DISCIPLINARY SOCIETY: 1981

	Percentage Who Have Served as . . .			
	National Officer	Committee Chair	Committee Member	N
Total	38%	49%	63%	281
Sector				
For-profit	17	39	51	59
University	43	54	68	147
Nonprofit	45	49	66	67
Discipline				
Biological Sci.	48	60	79	42
Physical Sci.	46	60	74	70
Social Sci.	30	43	52	54
Engineering-Prof	36	46	61	56
Other	31	39	48	54
Region				
Northeast	28	35	54	80
North Central	37	63	77	51
South	46	51	60	85
West	40	54	66	65

had served as a member of one or more national committees for a professional or disciplinary group.

Sixty percent of the leaders were members of the American Association for the Advancement of Science (AAAS). Fourteen percent were members of the National Academy of Sciences, 11 percent were members of the American Academy of Arts and Sciences, and 6 percent were members of the National Academy of Engineering (see table 6). Leaders affiliated with for-profit organizations were slightly less likely to be members of these invitational leadership societies.

Working from these organizational bases, the leaders reported a wide array of efforts to influence both colleagues within the scientific community and persons in the general political community on science policy matters. Approximately two-thirds of the leaders had attempted to influence the policy views of their colleagues in their own discipline or profession, and about 60 percent had sought to influence colleagues from other disciplines and professions (see table 7). Approximately half of the leaders reported seeking to influence persons outside the scientific community who were interested in, and informed about, science and technology matters: the attentive public for science policy. One in five of the leaders had attempted to influence the attitudes of the general public on science and technology issues, with leaders from the university sector being the least likely to have attempted to influence the general public.

The development of consensus on science policy matters requires frequent

TABLE 6
MEMBERSHIP IN NATIONAL LEADERSHIP ORGANIZATIONS: 1981

	Percentage Who Are Members of . . .				
	AAAS	National Academy of Sciences	National Academy of Engineering	American Academy of Arts and Sciences	N
Total	60%	14%	6%	11%	281
Sector					
For-profit	34	10	5	5	59
University	71	16	6	15	147
Nonprofit	60	13	6	6	67
Discipline					
Biological Sci.	79	17	2	10	42
Physical Sci.	73	17	7	13	70
Social Sci.	67	6	0	13	54
Engineering-Prof	48	21	14	9	56
Other	39	11	2	9	54
Region					
Northeast	60	13	4	15	80
North Central	59	10	6	10	51
South	58	13	6	5	85
West	65	22	8	14	65

TABLE 7
EFFORTS TO INFLUENCE PUBLIC ATTITUDES ON SCIENCE POLICY: 1981

	Percentage Seeking to Influence Persons from . . .				
	Own Discipline	Other Discipline	Attentive Public	General Public	N
Total	65%	59%	51%	22%	281
Sector					
For-profit	61	55	48	22	59
University	64	60	50	17	147
Nonprofit	79	69	61	36	67
Discipline					
Biological Sci.	69	62	64	21	42
Physical Sci.	71	67	49	19	70
Social Sci.	61	59	41	20	54
Engineering-Prof	64	59	57	23	56
Other	61	50	50	26	54
Region					
Northeast	66	60	54	29	80
North Central	59	65	49	20	51
South	67	60	53	25	85
West	66	54	46	12	65

TABLE 8
EFFORTS TO INFLUENCE OTHER LEADERS ON SCIENCE POLICY: 1981

	Percentage Seeking to Influence Leaders from . . .			
	Professional-Disciplinary Societies	State & Local Gov't	Church and/or Community	N
Total	50%	49%	20%	281
Sector				
For-profit	44	48	12	59
University	50	50	18	147
Nonprofit	60	54	33	67
Discipline				
Biological Sci.	50	57	14	42
Physical Sci.	51	56	17	70
Social Sci.	44	39	22	54
Engineering-Prof	57	54	21	56
Other	46	43	20	54
Region				
Northeast	49	44	20	80
North Central	41	49	14	51
South	54	49	29	85
West	52	55	11	65

discussions both within the leadership group and with external political and community leaders. Half of the leaders reported that they had attempted to influence other leaders of scientific or professional societies with their own policy views (see table 8). The same proportion reported seeking to influence state or local governmental leaders on a science policy matter, but only 20 percent indicated that they had made a similar effort to influence the views of community leaders.

In summary, the nongovernmental leaders of science policy reflected a broad spectrum of disciplines and professions, were predominantly university based, and were predominantly male with a modal age in the early 50s. The leaders reported high levels of governance and leadership activities both within the scientific community and in regard to the broader political system. Appropriate to their leadership role, this group indicated a high level of persuasive activity.

THE ATTENTIVE PUBLIC FOR SCIENCE POLICY

Those citizens who are sufficiently interested in a given policy area to become and remain informed about that area are referred to as the attentive public for that issue area. When the policy leaders of science and technology make an appeal to "the public" for support in regard to a particular policy dispute, it is usually the attentive public to which the appeal actually is addressed. Although the attentive public includes only a segment of the total public, for matters of science policy it is the *de facto* public and the democratic check on policy leaders and decision makers.

In Almond's (1950) original conceptualization, inclusion in an attentive public required: (1) a high level of interest in an issue or area, (2) a functional level of knowledge about that issue or area, and (3) a pattern of regular information acquisition relevant to the issue or area. In the 1979 survey, interest was measured by a combination of direct inquiry (e.g., How interested are you in this issue?) and a set of hypothetical headlines concerning a wide range of issues. A factor analysis of the reported headline reading preferences produced a clear factor structure that allowed the identification of a set of headlines associated with an interest in science policy matters. These items were used to construct an index of interest, that was very strongly and positively associated with the responses to the direct inquiry measure of interest. Since both measures were strongly associated, the 1981 survey carried forward only the two direct inquiry items and used those items to measure interest.[6] In both 1979 and 1981, slightly over a third of the adult population expressed a high level of interest in issues concerning new scientific discoveries and issues concerning the use of new inventions and technologies (see table 9).

Women and men did not differ significantly in regard to their interest in "new scientific discoveries," but a significantly higher proportion of men expressed a high level of interest in "the use of new inventions and technologies." It would appear that women were not less likely to be interested in science, but rather were less interested in more applied and technological issues.

In both years, there was a significant and positive association between the level of formal education completed and the level of interest in both science and technology. This association was the strongest of all of the independent variables studied.

The definition of a functional level of knowledge about science and technology involved a two-step procedure. In both years, respondents were asked to rate themselves as very well informed, moderately well informed, or not very well informed on each of the same nine areas used to assess interest. Those respondents who reported that they were either very well or moderately well informed concerning either new scientific discoveries or the use of new inventions and technologies were classified as perceiving themselves to be functionally

[6] In the context of the interview, each respondent was asked to indicate whether he or she was very interested, somewhat interested, or not very interested in each of nine issues or areas. One of the issues concerned "new scientific discoveries" and a second involved the "use of new inventions and technologies."

TABLE 9
INTEREST IN SCIENCE AND TECHNOLOGY ISSUES

	Percentage "Very Interested" in Issues Concerning . . .					
	New Scientific Discoveries		Use of New Inventions and Technologies		N	
	1979	1981	1979	1981	1979	1981
Total	36%	38%	34%	34%	1635	3195
Age						
17–24	38	37	37	35	309	570
25–34	42	45	39	35	360	747
35–44	32	40	29	33	257	528
45–54	38	38	35	36	229	505
55–64	37	40	28	39	225	354
65 and over	27	28	30	25	241	488
Gender						
Female	35	36	27	28	854	1677
Male	37	41	41	40	768	1514
Education						
Less than high school	22	24	19	26	455	632
High school	39	38	36	33	931	1912
Baccalaureate	54	55	45	44	146	456
Graduate degree	53	48	58	42	92	192

knowledgeable.[7] The surveys indicated that about 60 percent of the adult population of the United States perceived themselves to be at least moderately well informed about science, technology, or both (see table 10).

Women were less likely than men to see themselves as knowledgeable about science and technology. As with interest, there was a strong and positive association between the level of formal education completed and the perceived level of knowledge about science and technology.

The first step in the definition of attentiveness to science and technology policy required that a respondent report a high level of interest in, and at least a moderate level of knowledge about, either the science or the technology area. In both 1979 and 1981, approximately 39 percent of the adult population met this minimum standard (see table 11).

The second step in the definition of attentiveness concerned a somewhat more rigorous test of knowledge about science and technology. In both years, respondents were asked to define the meaning of a "scientific study," and were

[7] The moderately well informed group was included since both surveys incorporated a battery of substantive information items which were used for a final determination of knowledgeability. In other studies where a substantive knowledge battery was not available, attentiveness has been calculated based on a high level of interest and a self-reported very well informed.

TABLE 10
PERCEIVED KNOWLEDGE ABOUT SCIENCE AND TECHNOLOGY ISSUES

	Percentage "Very Well" or "Moderately Well" Informed About Issues Concerning . . .					
	New Scientific Discoveries		Use of New Inventions and Technologies		N	
	1979	1981	1979	1981	1979	1981
Total	63%	63%	60%	61%	1635	3195
Age						
17–24	66	63	62	63	309	571
25–34	69	69	66	63	361	747
35–44	63	62	57	61	257	528
45–54	62	64	58	60	231	503
55–64	63	63	63	62	224	354
65 and over	50	57	53	54	239	488
Gender						
Female	59	58	54	54	853	1677
Male	68	69	67	68	768	1514
Education						
Less than high school	43	50	44	49	455	631
High school	67	63	64	60	929	1911
Baccalaureate	82	77	77	75	143	456
Graduate degree	87	80	82	69	92	192

TABLE 11
IDENTIFICATION OF THE ATTENTIVE AND POTENTIALLY ATTENTIVE PUBLICS

1979		1981
39.6%	Percentage reporting a high interest in science and at least a moderate level of science knowledge	39.3%
−18.8	Minus persons who scored less than two on the Index of Scientific Knowledge	−17.6
− 1.3	Minus persons who did not report a pattern of regular news acquisition	− 1.7
19.5	Attentive public for science policy	20.0
20.1	The interested public is a residual group with high interest but that did not pass the knowledge or information acquisition tests.	19.3

asked to rate their comprehension of three terms: radiation, DNA, and GNP.[8] An Index of Scientific Knowledge was constructed, scoring two points for a correct understanding of scientific study and one point each for a clear understanding of radiation, DNA, and GNP. Those respondents with a score of zero or one on the index were deleted from the attentiveness group (see table 11). This step resulted in a deletion of about 18 percent of the population from the potentially attentive group in both years.

The third and final step in the definition of attentiveness to science and technology policy was the certification of a regular pattern of relevant information acquisition. To be classified as attentive, it was necessary for a respondent to (1) read a newspaper daily or "almost every day," (2) read one or more news magazines "most of the time," (3) read one or more science magazines "most of the time," or (4) watch a television science show like *Nova* "most of the time." Respondents who failed to do any of the information acquisition activities were dropped from the attentive group, leaving an estimated 20 percent of the public attentive to science and technology policy in both years (see table 11).[9]

In the process of defining the attentive public for science and technology policy, a second relatively important residual group was identified, which will be referred to as the interested public for science policy. This group has a high level of interest in science and technology matters and perceives itself to be at least moderately well informed about science, technology, or both. The members of this group, however, were unable to demonstrate adequate substantive information to remain in the attentiveness classification. In both 1979 and 1981, about 20 percent of the adult population of the United States were members of the interested public for science policy (see table 11).

A comparison of the demographic distribution of the attentive public, the interested public, and the nonattentive public in 1979 and 1981 indicated that there was a very high degree of stability in the estimates across the two years. Because of the larger sample, the estimates from the 1981 survey should be taken as the more accurate (see table 12).

[8] In the interview, each respondent was asked to indicate whether he or she had a clear understanding of what it means to study something scientifically, a general sense of the meaning of the term, or little understanding of it. In both years, those respondents who asserted that they had a clear understanding of the term were asked to tell the interviewer what they thought it meant, "in your own words." The responses were recorded verbatim and were coded independently by two coders into a six-tier classification that ranged from an understanding of theory building and hypothesis testing to no understanding at all. For a more complete description of the coding procedures, see Miller (1982b).

[9] It should be noted that a previous technical report on the 1979 survey (Miller, Prewitt, and Pearson, 1980) utilized the more comprehensive 60-item definition and estimated that approximately 18 percent of the adult population was attentive to science and technology policy. To foster the maximum level of consistency for this analysis, the attentiveness rate for 1979 was reestimated using the new procedures and the revised estimate was 19.5 percent. The difference between the two estimates is not significant at the .05 level.

TABLE 12
A DEMOGRAPHIC PROFILE OF THE ATTENTIVE PUBLIC FOR SCIENCE POLICY

	Attentive Public		Interested Public		Nonattentive Public	
	1979	1981	1979	1981	1979	1981
Age						
17–24	24%	18%	17%	19%	18%	18%
25–34	32	33	21	18	19	22
35–44	12	16	15	15	17	17
45–54	12	15	17	17	14	16
55–64	13	10	14	15	14	10
65 and over	7	8	17	16	17	18
Gender						
Female	35	40	55	54	58	57
Male	65	60	45	46	42	43
Education						
Less than high school	7	8	26	19	36	24
High school	59	48	63	69	55	61
Baccalaureate	19	32	8	9	6	10
Graduate degree	16	12	4	3	3	5
College science courses						
None	46	NA	79	NA	82	NA
Some	54	NA	21	NA	18	NA
N =	319	637	328	617	988	1940

To assess the influence of education, age, gender, and exposure to a college-level science course, a multivariate logit analysis was performed on the 1979 data, using the methods described by Goodman (1978) and Fienberg (1980). Since this is the first logit utilized in this analysis, it may be useful to review the procedure. A logit analysis is analogous to a regressional analysis, with a dependent variable and two or more independent variables. For this analysis, attentiveness to science policy was treated as a dichotomous dependent variable, although the logit methodology could have utilized a polychotomous dependent variable. The level of formal education was dichotomized into those with and without a baccalaureate degree. The college science course variable was dichotomized into those persons who had completed one or more college-level science courses and those who had not. Age was grouped into three categories (17 to 34, 35 to 54, and 55 and over) and entered into the analysis as a trichotomous variable.

An examination of the distribution of the dependent variable within each combination of independent variables is useful in understanding the structure of multivariate cross-classification analysis. In regard to attentiveness to science policy, the multivariate distribution indicated that courses and gender were associated with attentiveness within all of the strata of the multivariate distribution

TABLE 13
MULTIVARIATE DISTRIBUTION OF ATTENTIVENESS: 1979

Education	Age	Gender	Science Courses	Attentive to Science Policy	N
No college	17–34	male	none	18%	199
			some	63	75
		female	none	14	228
			some	23	61
	35–54	male	none	12	171
			some	40	20
		female	none	7	190
			some	13	30
	55 and over	male	none	15	158
			some	—	12
		female	none	3	229
			some	—	19
College	17–34	male	none	—	7
			some	52	46
		female	none	—	6
			some	46	46
	35–54	male	none	—	7
			some	49	45
		female	none	—	4
			some	25	24
	55 and over	male	none	—	5
			some	69	26
		female	none	—	6
			some	—	16

(see table 13). By systematically comparing male and female respondents with similar characteristics, it is possible to determine that a higher proportion of males were attentive to science policy within each stratum, but this cell-by-cell procedure does not produce a single estimate of the relative influence of gender on the likelihood of attentiveness.

Whereas a traditional regression model seeks to place individual scores in a space defined by the dependent and independent variables, logit analysis estimates an expected frequency for each cell based on the marginal distributions of the variables, controlling for all relationships among the independent variables. This first estimate is referred to as the independence model since it would fit the data perfectly if there were no relationships between any of the independent variables and the dependent variable. The difference between the estimated cell frequencies and the actual data distribution is measured in terms of likelihood-ratio chi-squares[10] and, for this specific analysis, was 272.01 (see table 14). This sum is a measure of the total level of association between the independent variables and the dependent variable, and it is referred to as the total mutual dependence in the model. Goodman (1978) suggests that mutual dependence is analogous to variance in a traditional regression model, but that mutual dependence is a more appropriate term since it reflects the cell basis of the analysis.

In this specific analysis, a stepwise logit procedure was utilized. The first step added the direct effect of exposure to college-level science courses on attentiveness to the model, and the procedure produced a new set of estimated cell frequencies, utilizing the additional marginals inserted into the model. The resulting fit indicated that 125.42 chi-squares remained unexplained, but that 146.59 chi-squares of mutual dependence were accounted for by exposure to a college-level science course. The coefficient of multiple-partial determination (CMPD) is analogous to a multiple R^2 in a traditional regression, and indicated that approximately 54 percent of the total mutual dependence in the model was accounted for by the college-level science course variable (see table 14).

Following the stepwise procedure, the education variable was then added to the model. The resulting estimates indicated that the effect of a baccalaureate accounted for an additional 8 percent of the total mutual dependence. Substantively, this result meant that holding constant the level of exposure to college science courses, there was a residual effect of completing a baccalaureate on the likelihood of being attentive to science policy. This ''college effect'' has

TABLE 14
A LOGIT MODEL TO PREDICT ATTENTIVENESS TO SCIENCE POLICY: 1979

Models		df	LRX^2	CMPD
H1	CEAG,Y	23	272.01	—
H2	CEAG,YC	22	125.42	.539
	Difference due to YC	1	146.59	.539
H3	CEAG,YC,YE	21	103.33	.620
	Difference due to YE	1	22.099	.081
H4	CEAG,YC,YE,YA	19	81.42	.701
	Difference due to YA	2	21.91	.081
H5	CEAG,YC,YE,YA,YG	18	42.61	.843
	Difference due to YG	1	38.81	.143

Y = attentiveness
C = college science courses
E = education
A = age
G = gender

df = degrees of freedom
LRX^2 = likelihood ratio chi-square
CMPD = coefficient of multiple-partial determination

[10] Goodman (1978) recommends the use of likelihood-ratio chi-squares because they are based on a geometric mean and are thus additive across the entire model. In contrast, Pearson's chi-square is based on an arithmetic mean and would not be additive across larger and more complex models.

been noted by Newcomb (1976), Astin (1977), and others; but these data suggest that its magnitude is only about one-seventh as large as the impact of a college science course. The cumulative CMPD indicated that approximately 62 percent of the total mutual dependence in the model was accounted for by the combined effects of a college-level science course and the completion of a baccalaureate (see table 14).

Next, the age variable was added to the model and the results indicated that another 8 percent of the total mutual dependence was attributable to age. Substantively, this result means that there was a significant residual age effect on the likelihood of attentiveness to science policy, holding constant differences in formal education and exposure to a college-level science course. The influence of the age variable points to a period effect. Persons who matured to adulthood in the scientific environment of the post-1945 years were more likely to be attentive to science than persons socialized in earlier, less scientific, periods.

Finally, gender was added to the model and the resulting estimates indicated that there was a significant residual gender effect on attentiveness, even after taking education, science courses, and age into account. The logit model indicated that the residual effect of gender accounted for about 14 percent of the total mutual dependence in the model (see table 14). This result suggests that there has been significant sex-role socialization in American society that identified science and technology as male domains and that this socialization persisted even when education, science course exposure, and age were all held constant. In relative terms, the magnitude of this residual gender effect was larger than either the residual effect of age or education, suggesting that the impact of early sex-role socialization has been substantial.

The combination of all four independent variables in the logit model is called the "main-effects model," and the combined effects of the four independent variables accounted for 84 percent of the total mutual dependence. This was a very good fit.

In summary, an attentive public for science policy has been defined empirically and includes one in five American adults. Another 20 percent of the adult population were classified as "interested" in science policy matters, but were deficient in substantive information about science and technology. The remaining 60 percent of the population has little interest in, and a relatively low level of information about, science policy issues. Exposure to a college science course and the completion of a baccalaureate program were positively associated with attentiveness, as were youth (under age 35) and the male gender.

THE NONATTENTIVE PUBLIC

At the base of the policy pyramid is the nonattentive public, which has little interest in, or knowledge about, science policy. While some of these individuals may have little interest in public affairs generally, others may be attentive to agricultural policy, foreign policy, economic policy, local politics, or other current issues. Although the nonattentive public was somewhat older and less likely to hold a college degree than the attentive public for science policy, its demographic profile was not significantly different than that of the interested public for science policy in either 1979 or 1981 (see table 12).

Under normal circumstances, the nonattentive public would have no involvement in the formulation of science policy. When decision makers or policy leaders seek broader public support for a particular policy outcome, it is usually the attentive public to which their appeals are directed. The normative standards of the science policy leadership group would view appeals to persons who were uninterested in and, more importantly, unknowledgeable about science and technology as inappropriate leadership behavior.

There are, however, two instances in which the nonattentive public may have a significant impact on the formulation of science policy. First, when science policy questions are included in a formal referendum, the balance of power swings to the nonattentive public. Referenda on issues like nuclear power, fluoridation, and laetrile illustrate the potential involvement of the nonattentive public in science policy matters. Second, if a science policy matter were to become an electoral issue on which candidates for public offices were to take opposite positions and to campaign on these differences, then the balance of power would again swing to the nonattentive public, which comprises the majority of the electorate. To date, there is no record of any candidate for public office winning or losing on a science policy issue, but if differences on science policy matters were to become partisan issues, this would be a potential avenue for broader policy involvement by the nonattentive public.

While extensive participation by the nonattentive public in the formulation of science policy appears unlikely, it is important to recognize that the mass public always holds a final veto on all public policy matters if, and when, they become sufficiently interested or concerned. Like foreign policy, science policy in the postwar years has been a bipartisan low-profile activity, which has insured control of the policy process by the leadership group primarily and the attentive public secondarily. The prospects for a continuation of this approach will be discussed in the final chapter.

Given the lack of interest in, and knowledge about, science policy on the part of the nonattentive public, it is useful to specify what this segment of the population thinks about organized science. Hennessy (1972) has suggested a useful differentiation between opinions and attitudes. In Hennessy's terms, opinions reflect the general disposition of an individual toward an attitude object about which he has little interest or knowledge. Opinions are fragile and highly unstable. In contrast, attitudes refer to more deeply rooted views about a policy or issue that are based on the values and life goals of the individual. There would be a higher level of internal consistency among an individual's attitudes about a salient issue or policy and a greater degree of stability over time.

In terms of the stratified policy-formulation model described above, it would be expected that the members of an attentive public would have well-developed attitudes toward the issue cluster to which they were attentive. These attitudes would tend to be ordered and consistent over time. To study public participation in the formulation of policy in any given area, it would be necessary to understand the attitudes of the appropriate attentive public. In contrast, it would be unlikely to find many individuals in the nonattentive public with well-developed attitudes on issue areas to which they were not attentive, but these individuals might be expected to have a set of general dispositions toward various issue areas.

GENERAL DISPOSITIONS TOWARD SCIENCE AND TECHNOLOGY

As a preface to an examination of direct policy participation by the policy leaders and the attentive public for science policy, it is useful to focus briefly on the general dispositions of all strata of the system toward organized science. In broad strokes, it appears that the American public holds the previous achievements of science in high regard and expects even greater achievements in the future. When asked about the sources of U.S. influence in the world, there was broad agreement that ''our'' technological know-how and form of government were the two most important influences (see table 15). The nonattentive public ranked the American economic system third, in contrast to the attentive and interested publics, which ranked scientific creativity third. In the view of the nonattentive public, ''our'' scientific creativity and ''our'' educational system were tied for sixth.

TABLE 15
SOURCES OF U.S. INFLUENCE IN THE WORLD: 1979

Sources of U.S. Influence	Attentive Public	Interested Public	Nonattentive Public
Our technological know-how	68%	47%	40%
Our form of government	37	40	45
Our scientific creativity	33	29	17
Our economic system	31	27	26
Our natural resources	12	16	23
The racial and ethnic mixture of our population	7	11	13
Our religious heritage	6	15	19
Our educational system	6	15	17
N =	321	334	1595

Beyond this positive assessment of previous contributions to national influence, the public appears to hold moderate to high expectations for the future. When asked to estimate the likelihood that scientific research would be able to achieve selected objectives within the next 25 years, the attentive and interested publics were significantly more optimistic than the nonattentive public (see table 16). A majority of the attentive and interested publics thought it "very likely" that science would develop accurate earthquake predictions, desalinate seawater, and find a cure for the common forms of cancer in the next 25 years. Reflecting their shallower knowledge base, respondents who were not attentive to science policy were more reluctant to use the "very likely" response and were more disposed toward the "possible but not likely" position. All strata of the public thought it relatively unlikely that scientists and social scientists would be able to place communities in outer space, reduce crime, or control inflation in the next 25 years.

All strata of the public attributed substantial responsibility for the national standard of living to science and technology, but there was also evidence of an awareness of potential negative results. Approximately 90 percent of all strata of the public agreed that scientific discoveries and technological know-how were responsible for our standard of living and were making our lives healthier, easier, and more comfortable (see table 17). At the same time, a majority of the interested and nonattentive publics were willing to agree that scientific discoveries and new technologies were making "our lives change too fast." Over 40 percent

TABLE 16
PUBLIC EXPECTATIONS FOR FUTURE SCIENTIFIC ACHIEVEMENTS

	Attentive Public		Interested Public		Nonattentive Public	
	1979	1981	1979	1981	1979	1981
Percentage expecting science within the next 25 years to . . .						
Predict earthquakes*	72%	63%	54%	63%	46%	NA
Supply cheap energy*	81	74	60	65	50	NA
Cure cancer*	58	59	48	61	43	NA
Put communities in space*	28	21	18	23	13	NA
Reduce crime	14	NA	17	NA	14	NA
Desalinate seawater*	64	63	47	63	39	NA
Control inflation	NA	20	NA	28	NA	NA
Mean score on Index of Scientific Expectations (range: 0–5)	3.0	2.8	2.2	2.7	1.8	NA
N =	289	637	292	617	839	1940

* = included in Index of Scientific Expectations

TABLE 17
GENERAL DISPOSITIONS TOWARD SCIENCE AND TECHNOLOGY: 1979

Percentage Agreeing That . . .	Attentive Public	Interested Public	Nonattentive Public
Scientific discoveries are largely responsible for our standard of living in the U.S.	92%	90%	87%
Scientific discoveries are making our lives healthier, easier, and more comfortable	90	84	82
Technological know-how is largely responsible for our standard of living in the U.S.	96	91	90
Scientific discoveries make our lives change too fast	42	53	59
New technologies based on scientific discoveries make our lives change too fast	45	62	60
Scientific discoveries tend to break down people's ideas of right and wrong	27	43	43
Future scientific research is likely to cause more problems than to find solutions to our problems	18	35	41
N =	321	334	1595

of the interested and nonattentive publics concurred with the idea that scientific discoveries tend to break down people's ideas of right and wrong. In addition, 40 percent of the nonattentive public agreed that future scientific research is more likely to cause more problems than find solutions to current problems: an extremely pessimistic view.

These data suggest that a significant segment of the population has come to view science as a two-edged sword, capable of bringing significant advances in health and other material comforts but also challenging traditional views and values and perhaps creating new types of problems for society. When asked to assess this balance of risks and benefits, a substantial majority of all strata of the public concluded that the benefits of scientific research have outweighed actual and potential hazards (see table 18). In both 1979 and 1981, about three-quarters of the interested public and two-thirds of the nonattentive public con-

TABLE 18
PUBLIC PERCEPTION OF THE RELATIVE BENEFITS AND RISKS OF SCIENTIFIC RESEARCH

	Benefits Greater Than Risks			
	1979 %	1981 %	1979 N	1981 N
Attentive Public	88	90	322	637
Interested Public	74	79	335	617
Nonattentive Public	63	66	978	924

TABLE 19
MULTIVARIATE DISTRIBUTION OF RISK-BENEFIT ATTITUDES: 1979

Gender	Education	Science Courses	Attentive to Science	Benefits Greater Than Risks	N
Male	no college	none	yes	91%	79
			no	64	450
		some	yes	93	58
			no	80	51
	college	none	yes	—	8
			no	—	11
		some	yes	92	63
			no	83	54
Female	no college	none	yes	81	53
			no	61	594
		some	yes	81	21
			no	74	92
	college	none	yes	—	6
			no	—	9
		some	yes	84	32
			no	83	54

cluded that the benefits of organized science outweighed its risks. Nine of 10 science policy attentives reached the same conclusion.

While some individuals perceived various actual and potential negative effects from scientific research, a majority in all strata believed that the benefits outweighed or justified the risks (see table 19). This balanced view may be seen as healthy, since it allows an individual to accept and reconcile an instance of harm from scientific work without discarding his or her support for the larger scientific enterprise. Since almost a third of the nonattentive public rejected the conclusion that the benefits of scientific research were greater than its risks, it is important to better understand the structure of these reservations.

A multivariate logit analysis of the public's assessment of the risks and benefits of scientific research was conducted, following the same procedures utilized in the logit analyses reported previously. The dependent variable was dichotomized into those persons who concluded that the benefits were greater than the risks and those who did not share that view. The same dichotomies utilized in previous analyses were retained for college science course exposure, formal education, gender, and attentiveness. The percentage within each multivariate category that believed that the benefits outweighed the risks indicated that attentiveness to science policy and exposure to a college science course were strongly associated with a positive view of the risk-benefit ratio for scientific research.

The results of the multivariate logit analysis indicated that attentiveness to science policy and exposure to a college science course accounted for a substantial share of the total mutual dependence in the model. In a stepwise analysis, exposure to a college-level science course accounted for 43 percent of the total mutual dependence, while attentiveness to science policy explained an additional 31 percent (see table 20). Once exposure to a college science course was taken into account, the effects of the completion of a baccalaureate and gender were reduced to about 7 percent each: a minor contribution to the total model.

When this set of general dispositions is viewed as a composite, it appears that the public attributes to science and technology a central role in the nation's influence in the world and in the standard of living of the American people. The level of expectations for future scientific achievements indicates that the past is viewed as prologue, especially among the attentive and interested publics. The substantial proportion of the public that reported that the benefits of scientific research had outweighed its risks points to a solid foundation of public confidence in science and technology.

TABLE 20
A LOGIT MODEL TO PREDICT RISK-BENEFIT ASSESSMENT ATTITUDES: 1979

Models		df	LRX^2	CMPD
H1	CEAG,Y	15	111.91	—
H2	CEAG,YC	14	63.15	.436
	Difference due to YC	1	48.76	.436
H3	CEAG,YC,YE	13	55.27	.506
	Difference due to YE	1	7.88	.070
H4	CEAG,YC,YE,YG	12	48.56	.566
	Difference due to YG	1	6.71	.060
H5	CEAG,YC,YE,YG,YA	11	14.48	.871
	Difference due to YA	1	34.08	.305

Y = risk-benefit assessment
C = college science courses
E = education
G = gender
A = attentiveness

df = degrees of freedom
LRX^2 = likelihood ratio chi-square
CMPD = coefficient of multiple-partial determination

There is, however, a significant level of awareness among the public of actual and potential hazards associated with science and technology. There was widespread awareness that the pace of change has been increasing and that science and technology were at least partly responsible for this development. Some people appeared to view new scientific discoveries as a challenge to moral beliefs, particularly among the interested and nonattentive publics.

All of these results must be set in the context of Hennessy's differentiation of opinions and attitudes. It is likely that the views expressed by the attentive public reflected a combination of interest, knowledge, and other political and social values. In contrast, the views of the nonattentive public appeared significantly less well organized. For example, when asked about the sources of national influence, the nonattentive public ranked technical know-how relatively high and assigned relatively low ranks to both scientific creativity and the educational system, reflecting a profound misunderstanding of the sources and nature of technical know-how. Similarly, when asked in the 1979 survey about the scientific or non-scientific basis of astrology, almost half of the nonattentive public expressed the view that astrology was "very" or "partially" scientific. Given these indicators of confusion, it would be inappropriate to interpret the preceding data as reflecting firm and consistent attitudes on the part of the nonattentive public. These are fragile views at best and when confronted with a concrete issue in a referendum, for example, the endorsements of prominent entertainers and other public figures may be as important as the policy pronouncements of scientists, engineers, and scientific organizations. The major value of these data may be that they do not support the claim that there is a strong or growing public distrust of science in the United States or the beginning of an overt antiscience mood.

Chapter 5
The Science Policy Agenda

Under normal conditions, the nongovernmental leaders for science and technology policy will define the science policy agenda, present their proposed policies and actions to the decision makers, and lobby on behalf of those policies. This activity consumes most of the efforts of the officers and staff of the major disciplinary societies, professional associations, and science policy groups. It is now appropriate to ask what are the major items on the science policy agenda in the view of these leaders, and what are their substantive attitudes toward those issues.

The 1981 leadership survey asked each leader to rate the importance of each of several frequently mentioned science policy issues. For each issue, the survey asked each leader to classify the problem as "a major problem, a problem but not major, or not really a problem." The problem most often cited as "major" was the level of funding for basic scientific research. Two-thirds of the leaders labeled the level of scientific funding as a major problem. The proportion that considered the level of basic research funding to be a major issue was highest among leaders from the university sector and among biological and social scientists. A majority of the leaders in all sectors labeled the basic research funding issue as major (see table 21).

The inadequate level of public understanding of science was the second most frequently cited issue by the leadership group, with approximately half of the leaders assessing the issue to be of major importance. Leaders from the for-profit sector were somewhat less likely to view the level of public understanding of science as a major issue than were leaders from other sectors (see table 21).

The quality of precollegiate science education was the third most frequently cited issue. Almost half of the leaders labeled the quality of precollegiate science education to be a major problem. As with the public understanding of science, leaders from the for-profit sector were signficantly less likely to view science education as a major problem.

Approximately a third of the leaders indicated that the obsolescence of scientific instrumentation for research was a major concern. This problem was rated as major by 41 percent of the leaders from universities, indicating the sector in which the obsolescence question was felt most strongly. Leaders with backgrounds in the physical sciences were somewhat more likely to view the instrumentation issue as being of major importance.

TABLE 21
MAJOR PROBLEMS FOR SCIENCE AND TECHNOLOGY: 1979

	Percentage Classifying Following Problems as "Major"							
	A	B	C	D	E	F	G	N
Total	20%	24%	28%	34%	47%	48%	67%	281
Sector								
For-profit	17	33	20	26	38	32	54	59
University	20	18	31	41	49	52	77	147
Nonprofit	25	31	29	28	45	52	58	67
Discipline								
Biological Sci.	14	25	22	27	44	57	76	42
Physical Sci.	17	22	31	43	56	51	63	70
Social Sci.	29	24	28	31	47	45	76	54
Engineering-Prof	16	28	19	37	43	51	63	56
Other	23	22	35	32	44	40	64	54

A = failure to apply new scientific knowledge toward end products and uses
B = inadequate incentives for industrial research and development
C = inadequate training and research opportunities for young scientists
D = obsolescence of scientific instrumentation for research
E = the quality of precollegiate science education
F = inadequate public understanding of science
G = lower funding for basic scientific research

The availability of training and research opportunities for young scientists was felt to be a major issue by only 28 percent of the leadership group. Leaders with high levels of scientific leadership and policy leadership activities were more likely to cite the "opportunity for young scientists" issue as being of major importance; but, even in this stratum, only a third of the leaders were willing to classify the issue as major.

Fewer than a quarter of the leadership group indicated that they viewed either the incentives for industrial research and development or the conversion of scientific knowledge into end products to be major policy issues. Even among the leaders from the for-profit sector there was limited enthusiasm for either of these two issues as major science policy problems. A third of the leaders from the for-profit sector labeled the level of incentives for industrial research and development to be a major science policy issue, but only 17 percent were willing to define the conversion of scientific knowledge into end products to be a major issue (see table 21).

In summary, the nongovernmental leaders of American science and technology defined the major science policy issues to be (in order of importance): the level of funding for basic research, the level of public understanding of science and technology, the quality of precollegiate science education, the obsolescence of research instrumentation, the availability of research and training opportunities for young scientists, the level of incentives for industrial research and development, and the failure to apply new scientific knowledge to end products and uses.

THE DYNAMICS OF SCIENCE POLICY

In chapter 3, a conceptual model of public policy formulation for low-saliency issues was prescribed and the roles of the strata of the system were described. Implicit in the agenda-setting role described above is an acceptance and understanding of the basic functions of the stratified model. It is appropriate to ask if there is any evidence that the leaders and the attentive public recognize their differential roles in the formulation of science policy.

On the leadership side, the evidence indicates that most policy leaders recognize the role of the attentive public in the policy-formulation process. When asked whether or not the "interested and informed citizen can often have some influence on science policy if he is willing to make the effort," 80 percent of the leaders agreed that this attentive group could have a voice in the policy process (see table 22). The level of recognition of the role of the attentive public was uniformly high across sectors and disciplines.

On the other side, the evidence indicates that most of the attentive public is aware of its role in the policy-formulation process and actively pursues that role within the political system. When asked to agree or disagree with the same statement concerning the potential influence of the interested and informed citizen, 80 percent of the attentive public agreed with the proposition (see table 22). Approximately the same level of agreement was reported by the interested public. It should be noted that respondents in the interested public classified themselves as at least moderately well informed about science or technology matters, but that they subsequently were unable to demonstrate an adequate

TABLE 22
PERCEIVED INFLUENCE OF THE ATTENTIVE PUBLIC FOR SCIENCE POLICY: 1981

	The interested and informed citizen can often have some influence on science policy if he or she is willing to make the effort.				
	Strongly Agree	Agree	Disagree	Strongly Disagree	N
Policy leaders	16%	64%	17%	3%	281
Attentive public	19	61	17	3	634
Interested public	20	64	14	2	616
Nonattentive public	NA	NA	NA	NA	NA

level of knowledge to be categorized as attentive to science policy. It is likely that a substantial portion of the interested public think of themselves as informed and, thus, may have viewed the statement as applying to themselves.

Beyond an attitudinal recognition of their role, the overt political behavior of the attentive public appears to fit the activist model suggested by Almond (1950) in his original conceptualization and the subsequent amplifications of this model by Rosenau (1974). Almost three-quarters of the attentive public reported activity in one or more voluntary associations (see table 23). While the proportion of attentives active in church and religious groups did not differ significantly from the interested and nonattentive publics, the proportion of the attentive public that reported an active role in a business or professional association (40 percent) or in a political organization (10 percent) was significantly higher than the other segments of the public.

As Almond and Verba noted in *The Civic Culture* (1963), organizational activity is the foundation for political activity and the willingness to participate in voluntary organizations appears to be based on a more fundamental trust in people. Using the same set of items employed by Rosenberg (1957) and Almond

TABLE 23
SOCIAL AND POLITICAL BEHAVIOR OF THE ATTENTIVE PUBLIC

	Attentive Public	Interested Public	Nonattentive Public
Active member in . . . (1979)			
No voluntary organizations	27%	33%	37%
Church or religious group	40	38	44
Professional or business association	37	14	13
Labor union	7	10	9
Agricultural association	4	3	4
Political organization	10	5	4
Local civic association	17	16	15
Service club or fraternal organization	29	20	12
Art, drama, or sports group	32	21	16
Social Trust Index (1979)			
0 (low)	20	32	32
1	21	26	25
2	26	22	22
3 (high)	32	20	21
Contacted public official on policy-related matter (1981)	38	20	16
Contacted public official on science policy matter (1981)	6	3	1
N (1979) =	319	328	988
N (1981) =	311	305	925

and Verba (1963) to measure social trust, the 1979 survey results indicated that the attentive public held a significantly more trusting attitude toward other people than did either the interested or the nonattentive publics (see table 23).

This same disposition toward participation was reflected in the reported contacts by citizens with public officials in regard to policy issues. Almost 40 percent of the attentive public reported contacting a public official on some policy related matter in the preceding year, and 6 percent reported contacts on science policy matters (see table 23). In contrast, only 20 percent of the interested public reported any contact with a public official on a policy issue, and only 3 percent reported a contact on a science policy matter. This differential suggests that the higher level of social trust reported by attentives and their stronger knowledge base combined to generate the necessary confidence to make a policy contact and argue for a particular policy outcome.

The attentive public's acceptance and exercise of its role in the political system can also be seen in the context of the political-specialization typology discussed above. In both 1979 and 1981, a significantly higher proportion of the attentive public were issue-oriented activists than either the interested or nonattentive publics (see table 24). At the other end of the spectrum, the attentive public was significantly less likely to be politically inactive than the other segments of the public.

In summary, it would appear that the stratified model for the formulation of science policy provides an accurate description for normal times. The policy agenda formulated by the leadership group provides the framework within which science policy develops. When disagreements arise among science policy leaders themselves or between the policy leaders and the decision makers, efforts may be made to mobilize the attentive public for science policy. Given the role of the leaders in shaping the science policy agenda, it seems appropriate to structure the following analyses of the specific policy views of the leadership group and the attentive public along the lines of this policy agenda.

TABLE 24
LEVELS OF POLITICAL PARTICIPATION

	Attentive Public		Interested Public		Nonattentive Public	
	1979	1981	1979	1981	1979	1981
Inactive citizens	12%	5%	22%	12%	26%	18%
Ritual voters	0	0	13	9	24	17
Issue voters	20	14	13	15	8	9
Ritual actives	0	0	20	18	24	23
Single-issue actives	24	14	13	18	9	15
Multi-issue actives	44	68	20	29	9	17
N =	319	311	328	305	988	924

Chapter 6
The Acquisition of Resources

During the last three decades, federal spending for scientific research and development generally and for basic scientific research specifically increased (in constant dollar terms) until 1968 or 1969, then declined slightly for most of the decade of the 1970s (see figs. 1 and 2). In terms of real purchasing power, organized science has experienced a decade and a half without any real increase in resources. Not surprisingly, resource concerns dominated the current science policy agenda suggested by the leadership group. This chapter will examine the attitudes of both policy leaders and the attentive public on resource issues.

THE ADEQUACY OF SUPPORT FOR SCIENTIFIC RESEARCH

The first issue on the science policy agenda in the view of the nongovernmental leadership group was the adequacy of the present level of federal support for scientific research generally and for basic research in particular. The issue of research funding can be approached at both a general and a specific level. At the more general level, it will be recalled that Vannevar Bush argued in *The Endless Frontier* (1945) that one of the major benefits of continued federal support for scientific research and development would be a vigorous economy and international economic leadership. In 1981, each leader was asked whether he or she thought that the level of support for scientific research and development was "closely tied," "loosely tied," or "generally unrelated" to economic growth. Approximately 35 percent of the leaders thought that there was a close relationship between the level of support for scientific research and development and economic growth, while 59 percent classified the relationship as "loose" (see table 25). These data suggest that while only a minority of current science policy leaders still believe in a close relationship between the level of scientific research and the rate of economic growth, a substantial majority believed that scientific research does contribute toward economic growth.

Approximately 40 percent of the leadership group would like to see scientific research funding tied to a percentage of the Gross National Product (see table 25). Among those leaders favoring linking scientific research funding to

TABLE 25
THE RELATIONSHIP BETWEEN SCIENTIFIC RESEARCH AND ECONOMIC GROWTH: 1981

	Research & Growth Closely Tied	Tie Research Funding to Percentage of GNP	Mean Percentage Favored*	N
Total	35%	39%	4.3%	281
Sector				
For-profit	27	40	4.1	57
University	40	45	4.5	139
Nonprofit	28	26	4.1	65
Discipline				
Biological Sci.	55	50	4.8	40
Physical Sci.	40	49	4.2	68
Social Sci.	25	39	4.7	51
Engineering-Prof	33	34	3.5	53
Other	28	21	3.3	52

* Calculated only for those who favored tying research funding to a percentage of the Gross National Product.

the GNP, the mean percentage recommended was 4.3. This would be a substantial increase over the 1980 rate of 2.77 percent (NSB, 1981).

In general, leaders from the university sector were more likely to see a close relationship between scientific research funding and economic growth, more likely to favor a tie between the level of scientific research funding and the GNP, and willing to peg research funding at a higher percentage of GNP than leaders from other sectors. Leaders from the physical and biological sciences reported a similar set of policy beliefs and preferences.

In general, the attentive public for science policy appeared to assign a relatively high priority to funding for scientific research. When asked whether the government was spending too much, too little, or about the right amount on each of 11 program areas, half of the attentive public for science policy favored increased spending for scientific research. In contrast, only 36 percent of the interested public for science policy and 26 percent of the nonattentive public supported increased funding for scientific research (see table 26). A majority of the attentive public for science policy also supported increased spending for reducing the crime rate, helping older people, improving education and health care, controlling pollution, and providing and conserving energy. As noted above, the members of the attentive public for science policy tend to follow multiple issues, and these results would suggest that the conduct of scientific research was but one of several important objectives.

When viewed in the context of recent federal budget cuts for domestic spending, the relative priority assigned to scientific research by policy leaders

TABLE 26
PERCENTAGE SUPPORTING INCREASED SPENDING FOR SELECTED PROGRAMS: 1981

	Attentive Public	Interested Public	Nonattentive Public
Reducing the crime rate	72%	82%	78%
Helping older people	77	74	73
Improving education	63	60	62
Improving health care	59	62	61
Reducing & controlling pollution	63	56	49
Providing & conserving energy	64	54	47
Helping low-income persons	42	46	46
Developing & improving national defense weapons	32	43	32
Conducting scientific research	50	36	26
Exploring space	40	24	10
Preventing & treating drug addiction	24	18	14
N =	317	298	963

TABLE 27
WILLINGNESS TO EXEMPT SCIENCE FROM BUDGET CUTS: 1981

	Willing to Exempt Science from Budget Cuts	N
Policy leaders	53%	281
Attentive public	30	627
Interested public	25	603
Nonattentive public	NA	NA

and the attentive public becomes clearer. In the 1981 study, both the leadership group and the attentive and interested publics for science policy were asked if scientific research should be exempted from proposed federal budgetary reductions. A majority of science policy leaders favored an exemption for scientific research, but only 30 percent of the attentive public for science policy took the same position (see table 27). To a large extent, this difference reflects the multi-issue orientation of most science policy attentives and the more exclusive science policy orientation of the leadership group. It is also likely that the attentive public was less acutely aware of the impact of a decade without a real increase in purchasing power for scientific research.

Beyond the aggregate level of funding, a second issue concerns the objectives or focus of scientific research and development. At a general level, the issue is often framed in terms of the share of funding for basic research versus the share for applied research. The 1981 study posed a situation in which the overall level of funding for scientific research and development might be stable for a period of several years and asked the leaders if, in that context, they would favor a reallocation of resources toward basic research, a reallocation in favor of applied research, or a maintenance of the present balance. Approximately half of the leadership group favored a reallocation to increase funding for basic research at the expense of applied work. About 35 percent of the leaders indicated that they would favor retaining the present balance between basic and applied, while only 15 percent suggested a shift in favor of applied research (see table 28).

The leadership support for a reallocation toward basic research was not shared by the attentive public for science policy. When asked to make the same judgment, only 28 percent of the attentives indicated that scientific research funds should be reallocated toward basic research (see table 28). More than a third of the attentive public favored a reallocation in favor of applied research efforts, in contrast to only 15 percent of the policy leaders. The interested public for science policy was more supportive of the present balance, a conservative response reflective of their lower level of information.

The leadership preference for basic research and the attentive public's support for applied research was reflected in a series of specific reallocation judgments. In the context of the same hypothetical period of stable aggregate funding for scientific research and development, both leaders and attentives were asked in 1981 to judge whether each of several specific research areas should receive an increased share of these resources, a decreased share, or their current share. A majority of policy leaders supported an increased share of science funding for basic biological research, and a plurality supported an increased share for human learning research and basic chemistry research (see table 29). In contrast, a majority of the attentive public favored increased support for disease-specific medical research and human-learning research. A plurality of the attentive public favored an increased share of science and technology funding for economics research, space exploration, basic physics research, and studies of human behavior.

The conflict may be more apparent than real. The attentive public reflected

TABLE 28
PREFERRED BALANCE BETWEEN BASIC AND APPLIED RESEARCH: 1981

	Increase Basic Research	Maintain Present Balance	Increase Applied Research	N
Policy leaders	51%	35%	15%	281
Attentive public	28	36	35	627
Interested public	28	47	25	603
Nonattentive public	NA	NA	NA	NA

TABLE 29
FUNDING PRIORITIES WITHIN SCIENCE AND TECHNOLOGY: 1981

	Policy Leaders		Attentive Public	
Research Area	Increased Share	Decreased Share	Increased Share	Decreased Share
Basic biology	57%	3%	44%	6%
Human learning	45	18	67	7
Basic chemistry	44	3	48	4
Engineering	42	12	46	4
Basic physics	39	12	47	7
Human behavior	37	33	39	23
Space exploration	36	26	47	17
Disease-specific medical research	36	19	82	2
Mathematics	24	17	32	15
Economics	19	40	48	14
Weapons	12	56	36	26
N =	287	287	632	632

the popular concern with personal health when it ranked disease-specific medical research first, and its eighth-place ranking of basic biological research may have reflected a lack of understanding of the relationship between basic and applied research. The reversal of these rankings by the policy leaders may have reflected well a shared concern about the improvement of personal health, but a different perception about the most efficacious approach to achieving that objective. The high level of leader-attentive agreement on an increased share for learning research may be seen also as a reflection of a common set of values.

There are other areas of disageement that appear to reflect differing perceptions of the likely results of research in various areas. For example, while both the leaders and the attentive public have a strong common interest in a healthy economy, only 19 percent of the leaders would support an increased share of science funding for economics research, while 48 percent of the attentive public supported that reallocation (see table 29). Similarly, while both the leaders and the attentive public favor the defense of the nation, a majority of the leaders favored a decreased share of science and technology funding for weapons-related research, while 36 percent of the attentive public favored an increased share for that purpose. As above, both the leaders and the attentive public appear to share common values, but the attentive public apparently thought that increased support for economics research or weapons development would result in a stronger economy or an improved national defense, while the leaders doubted those relationships. These differences may be characterized as questions of strategy rather than conflicts of purpose.

In summary, it appears that both science policy leaders and the attentive public for science policy share a general commitment to governmental funding for scientific research and development. The policy leaders would appear to place a somewhat higher priority on scientific research than the attentive public, which tended to view scientific research as one of several valued objectives. The attentive public did not support an exemption for scientific research from the broad federal budget cuts of 1981, as favored by science policy leaders. The science policy leadership group expressed a significantly stronger commitment to basic research than did the attentive public, but a review of the objectives reflected in those choices suggested that the gap was narrower than it first appeared.

THE QUALITY OF PRECOLLEGIATE SCIENCE EDUCATION

The second highest ranking resource issue on the science policy agenda of the leadership group was science and engineering education. Almost half of the leaders ranked the quality of precollegiate science education as being a problem of major importance.

Three-quarters of the science policy leaders disagreed with the statement that "the quality of science instruction in American high schools is on the rise" (see table 30). Half of the attentive public for science policy shared this concern. Only 37 percent of the interested public viewed the quality of precollegiate science education as a problem.

The international standing of American science and engineering education has become an issue of concern to science policy leaders. Spurred by comparisons of American and Japanese science and technology, there has been a growing volume of commentary on the educational systems of the two nations. There was almost universal agreement, 91 percent, among science policy leaders that the United States should seek to be the international leader in science and technology education; but only 61 percent of the leaders thought that the United

TABLE 30
THE QUALITY OF HIGH SCHOOL SCIENCE INSTRUCTION: 1981

	The quality of science instruction in American high schools is on the rise.				
	Strongly Agree	Agree	Disagree	Strongly Disagree	N
Policy leaders	1%	25%	50%	24%	281
Attentive public	8	43	42	8	587
Interested public	11	52	32	5	571
Nonattentive public	NA	NA	NA	NA	NA

States held a postion of international leadership in science and engineering education (see table 31). A higher proportion of leaders from engineering and other professions concluded that the United States was in a leadership position than did policy leaders from the physical, biological, or social sciences.

These concerns were reflected in the relative priority assigned to funding for science and engineering education by the policy leaders and the attentive public for science policy. In the context of the level of federal funding for science and technology, 58 percent of the leaders indicated that they would favor a reallocation of those resources to increase the share for science and engineering education (see table 32). This level of support was higher than any of the basic or applied research areas discussed in the preceding section. Similar proportions of the attentive and interested publics shared this commitment to increased funding for science and engineering education (see table 32).

In summary, most science policy leaders characterized the quality of precollegiate science education as a major problem and this concern was shared by about half of the attentive public for science policy. A solid majority of the

TABLE 31
INTERNATIONAL POSITION OF U.S. SCIENCE AND ENGINEERING EDUCATION: 1981

	U.S. Ahead in Science Education	U.S. Should Be Ahead in Science Education	N
Total	61%	91%	281
Sector			
For-profit	70	95	56
University	57	92	139
Nonprofit	60	86	65
Discipline			
Biological Sci.	56	98	41
Physical Sci.	55	97	66
Social Sci.	59	87	51
Engineering-Prof	70	87	56
Other	63	87	51

TABLE 32
FUNDING PRIORITY FOR SCIENCE AND ENGINEERING EDUCATION: 1981

	Decrease Funding	Hold Constant	Increase Funding	N
Policy leaders	10%	32%	58%	281
Attentive public	4	39	57	634
Interested public	4	43	53	613
Nonattentive public	NA	NA	NA	NA

science policy leadership group and the attentive public favored increased funding for science education, even if it was reallocated from research funding.

INDUSTRIAL RESEARCH AND DEVELOPMENT

While most of the resource issues on the science policy agenda concerned governmental support for scientific research or science education, there was some concern about the incentives for industrial research and development. Among science policy leaders, the issue of incentives for industiral research and development was the sixth highest ranking concern. A quarter of the total leadership group characterized the adequacy of incentives for industrial research and development as a major problem. A third of the leaders from the industrial sector identified this issue as a major concern (see table 21).

There was broad agreement among both policy leaders and the attentive public that American industry should invest more heavily in scientific research and development (see table 33). Over 90 percent of both leaders and attentives expressed this view, including the leaders from the industrial sector.

At the same time, there was general agreement with the proposition that the federal government should encourage this development with increased tax incentives (see table 34). Approximately three-quarters of both the leadership group and the attentive public supported the idea of increased federal tax incentives to stimulate industrial research and development.

From these data, it is clear that science policy leaders and the attentive public for science policy share a commitment to the expansion of industrial research and development. The difference in the sophistication of the leaders and attentives was illustrated in the issue of patent law modification. A majority of the attentive public agreed with the idea that a modification of the patent laws to extend the period of exclusive use would result in a major increase in technological innovation in the United States (see table 35). In contrast, 73 percent of science policy leaders rejected this proposition, reflecting a more

TABLE 33
ATTITUDES TOWARD INDUSTRIAL INVESTMENT IN SCIENTIFIC R&D: 1981

	American industry should invest more heavily in scientific research and development.				
	Strongly Agree	Agree	Disagree	Strongly Disagree	N
Policy leaders	38%	57%	5%	0	281
Attentive public	37	55	8	0	632
Interested public	30	60	10	0	611
Nonattentive public	NA	NA	NA	NA	NA

TABLE 34
ATTITUDES TOWARD TAX INCENTIVES FOR INDUSTRIAL R&D: 1981

	The federal government should provide larger tax incentives to increase industrial research and development.				
	Strongly Agree	Agree	Disagree	Strongly Disagree	N
Policy leaders	23%	49%	24%	3%	281
Attentive public	24	49	25	3	632
Interested public	20	51	24	5	612
Nonattentive public	NA	NA	NA	NA	NA

TABLE 35
ATTITUDES TOWARD PATENT LAW MODIFICATION: 1981

	If patent laws in the U.S. were modified to extend the period of exclusive use, we would see a major increase in technological innovation.				
	Strongly Agree	Agree	Disagree	Strongly Disagree	N
Policy leaders	2%	25%	63%	10%	281
Attentive public	11	50	36	3	608
Interested public	12	62	24	3	580
Nonattentive public	NA	NA	NA	NA	NA

sophisticated understanding of the sources of technological innovation and the impact of patent laws.

On the basis of these data, it would appear that there is a high level of consensus among science policy leaders on the importance of industrial research and development and the need for increased federal tax incentives to achieve that objective. The attentive public appeared to be in complete agreement with both the general objective and the need for tax incentives. Not surprisingly, the policy leadership group displayed a more sophisticated understanding of the nature of industrial research and the sources of technological innovation than the attentive public.

THE INTERNATIONAL STANDING OF AMERICAN SCIENCE AND TECHNOLOGY

The international standing of American science and technology is a result of several factors, including the level of resources available for facilities and for the conduct of scientific research. As noted in chapter 2, the European scientific community had dominated most disciplines prior to 1940, but one of the important effects of the war was the movement of a number of leading European scientists to the United States. The new laboratories and resources that flowed from the war effort and continued in the postwar years led to a period of American dominance in a wide range of scientific disciplines. The launching of Sputnik symbolized the Soviet challenge to the dominant position of American science. More recently, the emergence of the scientific communities of western Europe and Japan have marked new challenges to the leadership role of American science.

Soon after taking office as the president's science advisor, George Keyworth asserted that it was unreasonable for the United States to expect to be the world leader in all fields of science. He recommended that the United States focus its efforts on selected fields and recognize the primacy of other nations in other fields. This represents a significant change in postwar science policy.

The science policy leadership group appears to be divided on this issue. A majority of the leaders expressed the view that the United States is no longer the world leader in most fields of basic and applied science. Only 19 percent of the leaders believed that the United States was ahead of other nations in "almost all" areas of basic scientific research. In contrast, over half of the leaders expressed the belief that the United States ought to be the leader in almost all areas of basic science (see table 36). Substantial majorities of both biological and physical scientists believed that the United States ought to seek a broad leadership position in almost all areas of basic science. In regard to applied science and technology, only 9 percent of the leaders asserted that the United States was ahead of other nations in almost all areas of applied science. Thirty-eight percent of the leaders felt that the United States should seek a dominant position in applied science and technology.

In summary, a majority of the leaders of American science and technology

TABLE 36
INTERNATIONAL POSITION OF U.S. SCIENCE AND TECHNOLOGY: 1981

	U.S. Ahead in "Almost All" Areas		U.S. Should Lead in "Almost All" Areas		
	Basic	Applied	Basic	Applied	N
Total	19%	9%	51%	38%	281
Sector					
For-profit	20	11	53	44	55
University	20	7	56	41	140
Nonprofit	17	10	38	30	66
Discipline					
Biological Sci.	17	10	64	41	41
Physical Sci.	20	4	59	46	68
Social Sci.	22	10	42	32	50
Engineering-Prof	21	11	39	26	53
Other	14	13	54	46	52

believe that the United States is not currently in a leadership position in either basic or applied science. A majority of science policy leaders would favor an effort to return the United States to a leadership position in most areas of basic scientific research, and a plurality of about 40 percent would favor a similar effort to reestablish a position of international leadership for the United States in most areas of applied science and technology.

SUMMARY

The issues at the top of the science policy agenda, in the leadership view, primarily concerned resources. Reflecting more than a decade without a real increase in research support, science policy leaders placed major emphasis on the need for increased funding for scientific research and development generally and for basic research in particular. The attentive public for science policy appeared to be supportive of increased funding for scientific research but was interested primarily in the utilization of science and technology in the solution of problems in health, education, and similar areas. The attentive public appeared to view support for scientific research as one of several desirable governmental activities, but refused to accord it special status—including an exemption from federal budget reductions.

The condition of science education in the United States was a major concern for both science policy leaders and for the attentive public for science policy. Both leaders and attentives favored increased federal support for science education, even if the resources were reallocated from resources currently assigned to other scientific activities.

There was also agreement by both leaders and attentives that American industry should invest more heavily in research and development. Increased federal tax incentives were favored by both leaders and attentives to stimulate industrial investment in scientific research.

Recognizing that the level of resources is only one of several factors affecting the international standing of American science and technology, fewer than 20 percent of science policy leaders believed that the United States held a leadership position in almost all areas of basic and applied science. A majority of the leaders, however, felt that the United States should seek to regain a position of broad scientific and technical leadership.

Chapter 7
The Preservation of Independence

The preponderance of resource issues on the science policy agenda does not signal an abandonment by the scientific community of its commitment to freedom from external interference. Rather, the conflicts over independence during, and at the end, of the Second World War resulted in several basic agreements that assured the scientific community of public resources and a substantial degree of independence. Often referred to as the "contract" between organized science and society (Brooks, 1978; Nelkin, 1978), the stability of these agreements over the last four decades has significantly reduced the anxiety about independence that characterized the scientific community prior to 1940.

There were, however, some independence-related issues on the science policy agenda. The three primary areas of concern were the public understanding of science, governmental regulation of science and technology, and the creationist challenge to science education. This chapter will examine the attitudes of science policy leaders and the attentive public for science policy on these issues.

THE PUBLIC UNDERSTANDING OF SCIENCE

To a large extent, the independence of science depends on an adequate level of public understanding of the purposes and processes of science. Science policy leaders appeared to recognize the importance of public understanding, ranking it second among the issues of the science policy agenda (see table 21).

In the 1981 studies, a majority of both science policy leaders and the attentive public for science policy rejected the idea that there is a growing public distrust in science in the United States. A substantial minority of both groups, however, did express concern about public distrust of science (see table 37). The relatively even division of the leadership and attentive groups on this matter signals that it is a topic of current discussion and concern. This concern was reflected in a recent issue of *Daedalus*, that focused on scientific literacy.

TABLE 37
PERCEIVED PUBLIC DISTRUST OF SCIENCE AND TECHNOLOGY: 1981

	There is a growing distrust of science in the United States today.				
	Strongly Agree	Agree	Disagree	Strongly Disagree	N
Policy leaders	5%	41%	50%	4%	281
Attentive public	6	37	49	8	631
Interested public	11	41	39	9	604
Nonattentive public	NA	NA	NA	NA	NA

TABLE 38
ESTIMATED PUBLIC UNDERSTANDING OF SELECTED SCIENTIFIC CONCEPTS: 1981

	Mean Estimated Percentage Who Understand . . .			
	Scientific Study	Radiation	DNA	N
Policy Leaders	14%	22%	9%	281
Sector				
For-profit	10	22	5	59
University	15	21	11	147
Nonprofit	14	20	9	67
1979 Public Report	22	49	22	1635

Science policy leaders appeared to have a relatively low estimate of the public's knowledgeability about science. When asked in 1981 to estimate the percentage of the public that correctly understood the meaning of scientific study, radiation, and DNA, the mean percentage estimates by the leaders were significantly lower than the public's estimates of its own knowledge (see table 38). It should be noted, however, that when those respondents in the 1981 survey who estimated that they correctly understood the process of scientific study were asked to define it, only half could provide a minimally acceptable description or definition. The revised figure, 11 percent, is much closer to the leadership estimate. It is clear that science policy leaders have come to expect a relatively low level of public understanding and the leadership's perception of the public's knowledgeability has some empirical basis.

Miller (1983) developed an index of scientific literacy that incorporated three basic dimensions: an understanding of the purposes and process of scientific investigation, a minimal command of basic scientific constructs like radiation or DNA, and an awareness of some current science policy issues. Using this index, only 7 percent of the total adult population in the United States qualified as scientifically literate. Even more troubling, only a third of the attentive public for science policy met this standard. It would appear that the concerns of science policy leaders about the adequacy of the public's understanding of science is well founded.

In summary, science policy leaders were concerned about the level of public understanding of science in the United States, ranking it second on the science policy agenda. While most leaders and attentives rejected the idea of an increasing level of public distrust of science, a substantial minority of both leaders and attentives were concerned about the problem. It appears that science policy leaders have recognized the relatively low level of public knowledge about science, although the placement of the problem relatively high on the science policy agenda would suggest that many leaders do not find the situation satisfactory. The low levels of scientific literacy in the population generally, and especially among attentives, provides solid evidence for the leadership's concern over this issue.

GOVERNMENTAL REGULATION OF SCIENCE AND TECHNOLOGY

The most direct form of interference in the pursuit of science is governmental regulation of scientific and technological activities. To assess leadership concern over government regulatory activities, each leader in the 1981 study was asked to evaluate the level of regulation of a number of areas. The results indicated that a majority of the leaders found the current level of government regulation in regard to the development of new pharmaceutical products, the addition of chemical additives to food, the production and sale of pesticides, the use of human subjects in experiments, and basic scientific research to be "about right" (see table 39). In the case of the regulation of the construction of new nuclear power plants, a plurality of the leaders expressed the view that the present level of regulation was about right, but about a third of the leadership group thought that current regulations in this area were "too low."

In general, leaders from the for-profit sector were more likely to see the current level of regulations as too high and university based leaders were more likely to criticize the level of regulation as being too low. The sharpest break came in regard to the development of new pharmaceutical products, with 62 percent of the leaders from the for-profit sector expressing the view that the current level of regulation was too high, and 59 percent of leaders from the other two sectors concluding that they were about right (Miller and Prewitt, 1982).

Overall, these data suggest that the leaders of science policy do not view the present level of government regulation of science and technology to be a significant threat to the independence of scientific inquiry. While there is a higher level of reservation among leaders from the for-profit sector, these attitudes may

TABLE 39
LEADERSHIP ASSESSMENT OF GOVERNMENT REGULATION: 1981

	The Present Level of Government Regulation Is:		
Regulated Activity	Too Low	About Right	Too High
Use of human subjects	11%	70%	19%
Recombinant DNA experiments	11	68	21
Basic scientific research	4	68	28
Production and sale of pesticides	29	59	12
Chemical additives	31	56	13
New pharmaceutical products	10	53	37
Construction of nuclear power plants	34	41	26

N = 281

be rooted as much in a concern about the respective roles of government and business as in the independence of scientific inquiry.

While science policy leaders were indicating an acceptance of current governmental regulation of primarily technological activities, the attentive, interested, and nonattentive publics were expressing a willingness to restrain more basic scientific activities. In 1979 and 1981, respondents were asked to consider a set of scientific studies and to indicate whether they felt that scientists "should be allowed" to undertake work in that area. Even though the wording implied that it would be all right to propose limiting the areas in which scientists might work, a person sensitive to the independence issue might have been expected to resist the temptation to impose restrictions. Approximately half of the attentive public were willing to restrict the work of scientists concerning the creation of new life forms and to prohibit research that might develop techniques for parents to control the sex of a child at the time of conception (see table 40). Substantially higher proportions of the interested and nonattentive public indicated a willingness to restrain scientific work on "new life forms," a phrase which seemed to tap considerable anxiety in all segments of the public. On substantive areas other than new life forms and the determination of the sex of a child at the time of conception, the attentive public was significantly less likely to impose restraints than either the interested or the nonattentive publics. This result would suggest that referenda on the restriction of specific scientific work might attract a significant level of public support.

To understand the roots of this level of willingness to restrain scientific inquiry on selected topics, a multivariate analysis was conducted. For the purpose of this analysis, the Index of Scientific Restraint was dichotomized into those persons who would not have prohibited any of the studies and those who

TABLE 40
WILLINGNESS TO RESTRAIN SCIENTIFIC INQUIRY: 1981

	Attentive Public		Interested Public		Nonattentive Public	
	1979	1981	1979	1981	1979	1981
Percentage willing to restrict studies concerning . . .						
People to live to 100*	20%	20%	28%	24%	40%	NA
Weather control*	20	20	28	23	33	NA
Creation of new life forms*	56	51	69	67	78	NA
Intelligent beings in space*	12	24	36	27	51	NA
Criminal tendencies in young	14	NA	12	NA	19	NA
Selecting gender of child	NA	50	NA	57	NA	NA
Mean score on Index of Scientific Restraint (range: 0–4)	1.0	1.1	1.5	1.4	1.8	NA
N =	289	687	292	617	839	1940

* = included in Index of Scientific Restraint

would have restricted inquiry into one or more topics. Following the same procedures utilized in previous logit analyses, the level of education was dichotomized into college graduates and noncollege graduates, and exposure to a college-level science course was divided into those with no college science course exposure and those with some.[11] Attentiveness was also dichotomized, with the interested and the nonattentive publics grouped together. Gender was also included in the model.

The effects of these four independent variables on the willingness to restrain scientific inquiry can be seen by examining the percentage willing to restrain one or more areas of scientific research within each of the multivariate categories (see table 41). Attentiveness, gender, and exposure to a college science course all appear to have a significant impact on the proportion of individuals willing to support restriction.

A multivariate stepwise logit analysis indicated that gender accounted for approximately 39 percent of the total mutual dependence in the model, that exposure to a college-level science course accounted for 36 percent of the total mutual dependence, and that attentiveness explained 16 percent of the total

[11] While this variable is strongly associated with a baccalaureate degree, it is not an identical measure. Many respondents have taken college-level science courses without finishing a baccalaureate. This experience is particularly relevant to community and junior college students who have taken science courses as a part of a program to prepare for various technical occupations. Conversely, some college graduates report that they did not take a college-level science course during their undergraduate program.

TABLE 41
MULTIVARIATE DISTRIBUTION OF WILLINGNESS TO RESTRAIN SCIENTIFIC INQUIRY: 1979

Gender	Education	Science Courses	Attentive to Science	Willing to Restrain	N
Male	no college	none	yes	64%	78
			no	77	450
		some	yes	44	57
			no	59	51
	college	none	yes	—	8
			no	—	11
		some	yes	54	63
			no	70	54
Female	no college	none	yes	81	53
			no	88	594
		some	yes	75	20
			no	84	93
	college	none	yes	—	7
			no	—	10
		some	yes	56	32
			no	74	54

TABLE 42
A STEPWISE LOGIT MODEL TO PREDICT WILLINGNESS TO RESTRAIN SCIENTIFIC INQUIRY

Models		df	LRX^2	CMPD
H1	CEAG,Y	15	124.64	—
H2	CEAG,YC	14	79.36	.363
	Difference due to YC	1	45.28	.363
H3	CEAG,YC,YE	13	77.40	.379
	Difference due to YE	1	1.96	.016
H4	CEAG,YC,YE,YG	12	28.58	.771
	Difference due to YG	1	48.82	.392
H5	CEAG,YC,YE,YG,YA	11	8.71	.930
	Difference due to YA	1	19.87	.159

Y = willingness to restrain
C = college science courses
E = education
G = gender
A = attentiveness

df = degrees of freedom
LRX^2 = likelihood ratio chi-square
CMPD = coefficient of multiple-partial determination

mutual dependence (see table 42). The effect of a baccalaureate degree was not significant once gender, science course exposure, and attentiveness were held constant. The significantly higher proportion of women who were willing to restrain scientific inquiry, especially in regard to the creation of new life forms and the determination of gender, will be discussed in greater detail in the final chapter.

In summary, science policy leaders did not display any significant concern about the current level of governmental regulation of a wide array of scientific and technological activities. At the same time, the attentive, interested, and nonattentive publics all indicated a willingness to restrain various scientific activities, especially research concerning new life forms and increased control of gender determination. While the propensity of a substantial portion of the public to restrain scientific research has not become a widespread political problem for the scientific community, it remains a potential source of exploitation and difficulty.[12]

THE CREATIONIST CHALLENGE

Another dimension of independence concerns the intervention of the government or groups outside the scientific community in the teaching of science. The claim by various religious groups that the biblical version of creation should be taught in science classes on an equal footing with Darwin's theory of evolution represents a challenge to the authority of the scientific community to define the content of science education. Some states recognized this claim and enacted legislation requiring the teaching of creationism. The federal courts recently voided an Arkansas statute requiring the teaching of creationism.

Science policy leaders are virtually unanimous in their rejection of the creationist claim (see table 43). The opposition to creationism was uniformly high among leaders from all sectors and disciplines.

In sharp contrast, two-thirds of the attentive public for science policy agreed with the creationist position. This represents one of the sharpest and most critical differences found between science policy leaders and the attentive public for science policy. Given the clear wording of the statement and the direct reference to the "biblical version of creation," it cannot be assumed that very many respondents misunderstood the question. Some attentives may have been attracted to the policy of equal treatment of all ideas, but the inescapable conclusion of the data is that a substantial portion of American adults had reservations about the theory of evolution and were unaware or unconcerned about the issues

[12] The two major cases of overt attempts to restrain scientific work in recombinant DNA were efforts by the city governments of Cambridge, Massachusetts, and Madison, Wisconsin to prohibit the construction of P-3 research facilities in their jurisdictions.

TABLE 43
ATTITUDES TOWARD CREATIONISM: 1981

	The biblical version of creation should be given equal weight with the theory of evolution in the teaching of science in the public schools.				
	Strongly Agree	Agree	Disagree	Strongly Disagree	N
Policy leaders	2%	8%	31%	59%	281
Attentive public	28	39	18	15	630
Interested public	27	43	19	11	607
Nonattentive public	NA	NA	NA	NA	NA

of scientific method or evidence. In the view of science policy leaders, this result would be seen as compelling evidence for a major improvement in both science education and the public understanding of science.

SUMMARY

Looking at the full array of independence issues, it is necessary to conclude that most of the problems and solutions appear to be long term in character. The basic compact between organized science and the government has been observed by both sides over the last four decades, and the level of anxiety about external interference in the affairs of science appears to have diminished among science policy leaders. There is general acceptance among science policy leaders of current government regulation of science and technology.

At the same time, the science policy leadership appears to have recognized the low level of public understanding of science in the United States as a major long-term problem. The low level of scientific literacy among the attentive public is a serious barrier to communicating sophisticated information about science policy, or to effectively mobilizing the attentive public to work for specific policy outcomes. The propensity of the attentive, interested, and nonattentive publics to restrain scientific investigation in selected areas to support the creationist position in regard to science education illustrates the potential threats to the independence of the scientific community that may result from low levels of scientific literacy.

Chapter 8
Public Participation in Specific Controversies

In chapters 2 and 3, the concept of issue attentiveness was introduced and science and technology policy was defined as an object of attentiveness. This broad focus is useful and appropriate since a significant portion of federal legislative and executive actions concerns all of science and technology. At the same time, some individuals focus on narrower, more specific science-related controversies and issues. This chapter will examine attentiveness to three specific issues, looking at the relationship between specific issue attentiveness and: science policy, substantive policy attitudes, and the likelihood of personal participation in three hypothetical disputes.

SPECIFIC ISSUE ATTENTIVENESS

To examine the structure of public attitudes toward specific science-related issues and controversies, the analysis will focus on the chemical food additive controversy, the nuclear power controversy, and an issue concerning space exploration. The food additive issue was selected because it represents a direct choice by consumers concerning products that they consume as well as a legislative and executive policy issue. Nuclear power plant siting controversies have occurred across the nation and the disputes have involved numerous community disputes and several referenda. Although this issue had been selected for inclusion in the study prior to the accident at Three Mile Island (TMI), the data collection for the 1979 survey occurred about seven months after TMI and provides an opportunity to examine the structure of public attitudes on a science-related matter that had substantial national media attention in the months preceding the interview period. The space exploration issue was selected because it represented a controversy that did not touch the daily lives of respondents directly but reflected a mix of budgetary and other interesting policy questions. The combination of these three issues should provide a good overview of the structure and content of public attitudes on specific science-related controversies.

Following the general approach of the stratified model, it is reasonable to expect that there are segments of the public that are especially interested in, and informed about, each of these specific controversies, and these groups may be conceptualized as specific issue attentive publics. To qualify as attentive to any of the three specific controversies, a person would need to display a high level of interest in that issue, be reasonably knowledgeable about the substance of the controversy, and exhibit a pattern of relevant information acquisition. The 1979 survey was designed to examine specific issue attentiveness, substantive policy preferences, and probable personal participation in three hypothetical controversies. The data provide a rare opportunity to compare both specific issue attentiveness and general science policy attentiveness.

To qualify as interested in a specific issue, a respondent had to report that he or she was "very interested" in that particular issue and also had to report in a separate section that he or she would have read several hypothetical headlines relevant to that specific issue.[13] Using this approach, the data from the 1979 survey indicated that approximately 56 percent of the public were interested in the chemical food additive issue, 58 percent in the nuclear power issue, and 26 percent in controversies involving space exploration (see table 44).

Attentiveness, however, requires both a high level of interest and a functional level of knowledge about the substantive issue area. Each respondent in the 1979 survey was also asked to name two benefits that might be expected from food additives, nuclear power, and space exploration and two harms or dangers associated with each activity. To be classified as informed on any given issue, a respondent who classified himself as very well informed had to be able to name at least two benefits or harms for that issue. A respondent who classified himself as only moderately well informed had to name three or four benefits and harms. As a result of this procedure, those respondents who failed to qualify as knowledeable were dropped from the potential attentive public group.

A third requirement for attentiveness is a regular pattern of information

[13] The number of headlines related to each of the three areas differed, thus the number required to qualify as interested also differed. The 1979 survey included three headlines concerning chemical food additives and it was necessary for a respondent who classified himself as very interested to have indicated an intention to read at least two of the three headlines to be classified as interested for the attentiveness calculation. A respondent who classified himself as "somewhat interested" in the food additive issue but who indicated that he would have read all three headlines was also classified as interested in the food additive issue.

The 1979 survey included six headlines related to nuclear power and respondents who classified themselves as very interested in the issue needed to have indicated an intention to read three of those headlines to be classified as interested in that issue. Respondents who classified themselves as "somewhat interested" in the nuclear power issue but who would have read five or six of the headlines were also classified as interested.

Only two of the headlines in the 1979 study related directly to space exploration. Respondents who reported that they were very interested in the issue and who would have read at least one of the headlines were classified as interested, and respondents who were somewhat interested in the issue area but who would have read both of the headlines were also categorized as interested in that specific issue area.

TABLE 44
ATTENTIVE PUBLICS FOR SPECIFIC ISSUES: 1979

	Issue Area		
	Space Exploration	Nuclear Power	Food Additives
Interested in specific issue area	25.7%	58.4%	56.1%
Minus persons low in knowledge in issue area	−17.0	−30.6	−28.8
Minus persons not regular consumers of relevant information	− .3	− 1.9	− 1.2
Attentive public for specific issue	8.4	25.8	24.9
Interested public for specific issue	17.3	27.8	21.2
N =	1635	1635	1635

acquisition or consumption relevant to the substantive area. All three of the specific issues included in this analysis have had prominent and continuing media coverage. Accordingly, the measure of acquisition utilized in the determination of attentiveness to science policy generally was repeated for each of the three specific issue attentiveness calculations. It will be recalled that this measure required a respondent to report regular readership of a newspaper, or regular viewership of television news, or regular readership of a science-oriented magazine. As with the determination of attentiveness to general science policy, this criterion reduced the potential attentive population by one to two percentage points, reflecting the strong association of this measure with the level of substantive knowledge itself.

When the three criteria of interest, knowledge, and information acquisition were combined, the data from the 1979 survey indicated that approximately 8 percent of the adult population were attentive to space exploration, 26 percent were attentive to the nuclear power issue, and 25 percent were attentive to the chemical food additive controversy (see table 44). These same data indicated that approximately 17 percent of the adult population were included in an interested public for space exploration, 31 percent in an interested public for nuclear power, and 13 percent in an interested public for food additive issues.

An analysis of the composition of each of the three specific issue attentive publics revealed two significant differences. First, the attentive publics for space exploration and nuclear power were predominantly male, whereas the attentive public for food additive issues was predominantly female (see table 45). This difference points to an important facet of public attitudes toward science and technology policy. The range of scientific and technological activities is wide and impacts on a large number of personal or potential concerns of citizens.

The motivation for following a broad area like science policy may differ significantly among individuals, reflecting their own interest and value patterns. The finding that female respondents were more likely to be attentive to the food additive issue reflects in large part the responsibility of women for the purchase and preparation of food within most families. While the distinction between

TABLE 45
DEMOGRAPHIC PROFILE OF SPECIFIC ISSUE ATTENTIVE PUBLICS: 1979

	Attentive Public for . . .		
	Space Exploration	Nuclear Power	Food Additives
Total	8%	26%	25%
Age			
17–24	23	19	15
25–34	35	26	30
35–44	15	15	16
45–54	13	16	14
55–64	9	16	16
65 and over	5	9	9
Gender			
Female	29	37	59
Male	71	63	42
Education			
Less than high school	7	12	10
High school	66	60	61
Baccalaureate	15	16	18
Graduate degree	11	12	12
Attentiveness to Science Policy			
Attentive public	58	42	37
Interested public	20	19	20
Nonattentive public	22	39	43
Science Courses			
None	46	56	56
Some	55	44	44
N =	137	423	407

male and female roles within families is undoubtedly changing, most married women still exercise the major responsibility for food shopping and preparation, and women who are single or single parents bear the total responsibility for food selection. Many of the questions inherent in the food additive controversy are scientific in nature, thus, this is one route by which the responsibilities of women may lead them to an interest in science and a need to acquire relevant information. Conversely, the finding that the attentive publics for nuclear power and space exploration were predominantly male probably reflects the greater likelihood of males holding occupational roles in which these issues are relevant or reading materials in which these issues are discussed.

Second, a significantly higher proportion of the attentive public for space exploration had experienced a college science course than either of the other two specific issue attentive publics (see table 45). This result suggests that college science courses may stimulate attentiveness to activities such as space exploration that have minimal personal impact, while the potential direct effects of activities such as nuclear power and food additives may stimulate attentiveness to those issues.

To determine the relative contribution of these factors to the development of attentiveness to these three specific issues, a set of stepwise logit analyses was conducted. In each analysis, the dependent variable was dichotomized into those persons who were attentive to that specific issue and those who were not. Exposure to college science courses, formal education, and gender were dichotomized in the preceding logit analyses. Age was entered into the model as a trichotomous variable, reflecting the groupings 18-34, 35-54, and 55 and over.

The three stepwise logit analyses indicated that exposure to a college science course was the most important predictor of attentiveness for all three specific issues (see table 46). Exposure to a college science course was entered into the model as the first variable and accounted for about a third of the total mutual dependence in all three analyses. The second most influential variable for all three specific issues was attentiveness to science policy, which accounted for approximately 30 percent of the total mutual dependence in regard to space exploration and nuclear power and 20 percent in regard to food additives. This result is supportive of the conclusion that a major path to specific issue attentiveness comes through attentiveness to more general science policy matters. For both the space exploration and nuclear power siting issues, gender was the third most important variable, accounting for about 14 percent of the mutual dependence in both models. Gender was not significantly associated with attentiveness to the food additive issue, indicating that women had reached parity in that area. The third most important predictor of attentiveness to the food additive issue was the completion of a baccalaureate, that accounted for 13 percent of the total mutual dependence in that model.

It would be reasonable to expect a high degree of association between attentiveness to science policy generally and attentiveness to these three specific

TABLE 46
THREE LOGIT MODELS TO PREDICT ATTENTIVENESS TO SPECIFIC CONTROVERSIES: 1979

Models		Coefficient of Multiple-Partial Determination When Y Represents . . .		
		Space Exploration	Nuclear Power	Food Additives
H1	CEAGZ,YC	.300	.339	.342
	Difference due to YC	.300	.339	.342
H2	CEAGZ,YC,YE	.304	.400	.467
	Difference due to YE	.000	.060	.125
H3	CEAGZ,YC,YE,YG	.437	.553	.522
	Difference due to YG	.134	.152	.055
H4	CEAGZ,YC,YE,YG,YA	.488	.553	.527
	Difference due to YA	.051	.000	.005
H5	CEAGZ,YC,YE,YG,YA,YZ	.791	.859	.728
	Difference due to YZ	.303	.306	.201

Y = attentiveness to specific issues
C = college science courses
E = education
G = gender
A = age
Z = attentiveness to science policy

TABLE 47
GENERAL AND SPECIFIC ISSUE ATTENTIVENESS: 1979

Attentive Public for . . .	Percentage Attentive to . . . Science Policy	Space Exploration	Nuclear Power	Food Additives	N
Science policy	—	25	56	47	322
Space exploration	58	—	65	51	137
Nuclear power	42	21	—	46	423
Food additives	37	17	48	—	407

issues, and the data from the 1979 survey generally supported that expectation. Of those persons attentive to science policy generally, 56 percent were also attentive to the nuclear power issue, 47 percent were attentive to the food additive controversy, and 25 percent were attentive to space exploration (see table 47). A majority of those persons who were attentive to space exploration, the smallest attentive public of the four, were also attentive to science policy, nuclear power, and food additives. Approximately 40 percent of those who followed nuclear power were also attentive to science policy. A similar proportion of the nuclear power attentives followed the food additive controversy, but only 21 percent of this group were attentive to space exploration. Almost half of the food additive attentives also followed the nuclear power dispute, while 37 percent were attentive to science policy generally and 17 percent followed space exploration.

While the level of association among the four areas was apparent in the 1979 data, it was also clear that there were significant portions of these specific issue publics that were attentive to only a single controversy. These data suggest two possible paths to specific and general science policy attentiveness. First, a person who was already attentive to science policy issues generally would be more likely to hear about other science-related issues concerning more limited areas and to develop a personal interest and knowledge base about those specific matters. Given the number of specific issues that arise, it is unlikely that science policy attentives would elect to follow all of the controversies, thus there would be a second level of issue specialization within a broader issue area.

Conversely, some individuals may become attentive to one specific controversy and either remain attentive to only that area or become more aware of broader science policy matters and ultimately become attentive to general science policy. The large number of specific issue attentives who were not attentive to science policy at the time of the 1979 survey indicated that a substantial portion of specific issue attentives were not yet attentive to science policy generally. A determination of the proportion of individuals that develop general science policy attentiveness through attentiveness to one or more specific science-related controversies can only be determined through longitudinal studies and is beyond the capability of the data bases employed in this analysis.

In summary, it appears that there are separate and identifiable attentive publics for specific issues pertaining to science and technology. While there is a significant level of association among these different attentive publics, it is apparent that many individuals are attentive to only one or two specific controversies and not to science policy generally. Conversely, not all members of the attentive public for broader science policy matters concern themselves about all of the specific controversies within the boundaries of science and technology. The demographic differentiation among the three specific issue attentive publics and the attentive public for science policy reflected the variety of personal interests that can lead to attentiveness to an issue area.

SPECIFIC ISSUE POLICY VIEWS

The analysis in the preceding sections indicated that the attentive public for science policy tended to hold policy views more favorable to the scientific community than persons not attentive to science policy, and the policy views of the attentive public for science policy appeared to reflect a broader knowledge base and a more sophisticated understanding of the issues. It is appropriate to turn to the substantive policy views of each of these three specific issue attentive publics and inquire into the distribution and direction of their policy attitudes.

The most fundamental issue concerning space exploration is whether the

United States should continue to make a major investment of its scientific research and development resources in space exploration or whether those resources might be used better in other scientific or technological endeavors. When asked in the 1979 survey whether the space program should be continued, 60 percent of the adult population expressed a positive view (see table 48). Over 90 percent of those individuals attentive to space exploration favored a continuation of the program, and all of those persons attentive to both science policy and space exploration were favorable toward the continuation of the program. In contrast, only half of the public who were attentive to neither space nor science supported the continuation of the space program.

Although males, persons with some college science course exposure, and college graduates were all significantly more favorable to the space program than their counterparts, it will be recalled that all of these attributes were strongly and positively associated with attentiveness to space exploration. To identify the relative contribution of each of these components to the policy attitude toward space exploration, a stepwise logit analysis was conducted. The dependent variable was dichotomized into those who favored the continuation of

TABLE 48
POLICY PREFERENCE CONCERNING SPACE EXPLORATION: 1979

	Percentage in Favor of Space Exploration	N
Total	60%	1635
Attentive to . . .		
Space & science	100	79
Space only	90	58
Science only	84	243
Neither area	51	1255
Age		
17–24	68	309
25–34	70	361
35–44	60	258
45–54	65	234
55–64	51	228
65 and over	36	245
Gender		
Female	49	862
Male	71	773
Education		
Less than high school	37	465
High school	65	932
Baccalaureate	79	146
Graduate degree	85	92
College Science Courses		
None	53	1210
Some	78	425

the program and those who did not. The five independent variables were dichotomized as they have been in the preceding logit models. The results indicated that exposure to a college science course accounted for about 29 percent of the total mutual dependence and that gender and attentiveness to science policy each explained about 25 percent of the total mutual dependence (see table 49). Attentiveness to space exploration, that was added to the stepwise model last, accounted for 11 percent of the total mutual dependence. This result suggests that there was a marginal effect from specific issue attentiveness in the case of space exploration, but that the marginal effect was smaller than the effect of attentiveness to science policy itself.

The major policy issue in the nuclear power area has been the siting of specific nuclear power plants. Since this issue has inherently local dimensions, the 1979 survey of public attitudes toward science and technology asked each respondent to indicate whether he or she would favor the siting of a nuclear power plant in their own area. The results indicated that only 30 percent of the public would favor the siting of a nuclear power plant in their local area (see table 50). Individuals who were attentive to both science and nuclear power were the most likely to favor the placement of a nuclear power plant in their area, followed by those persons attentive to science policy but not to the nuclear power controversy itself. The public not attentive to either nuclear power or

TABLE 49
THREE LOGIT MODELS TO PREDICT POLICY PREFERENCES IN SPECIFIC CONTROVERSIES: 1979

		Coefficient of Multiple-Partial Determination When Y Represents . . .		
Models		Space Exploration	Nuclear Power	Food Additives
H1	AZGEC,YC	.290	.089	.031
	Difference due to YC	.290	.089	.031
H2	AZGEC,YC,YE	.322	.100	.031
	Difference due to YE	.032	.011	.000
H3	AZGEC,YC,YE,YG	.570	.638	.185
	Difference due to YG	.247	.539	.154
H4	AZGEC,YC,YE,YG,YZ	.815	.740	.187
	Difference due to YZ	.245	.101	.002
H5	AZGEC,YC,YE,YG,YZ,YA	.923	.758	.702
	Difference due to YA	.108	.019	.515

Y = policy preferences
C = college science courses
E = education
G = gender
Z = attentiveness to science policy
A = attentiveness to specific issue

TABLE 50
POLICY PREFERENCE CONCERNING NUCLEAR POWER: 1979

	Percentage in Favor of Locating Nuclear Power Plant in Area	N
Total	30%	1635
Attentive to . . .		
Nuclear power & science	48	179
Nuclear power only	34	244
Science only	41	143
Neither area	25	1069
Age		
17–24	24	309
25–34	24	361
35–44	38	258
45–54	37	234
55–64	33	228
65 and over	29	245
Gender		
Female	20	862
Male	42	773
Education		
Less than high school	24	465
High school	31	932
Baccalaureate	31	146
Graduate degree	49	92
College Science Courses		
None	28	1210
Some	37	425

science were the least receptive to a nuclear power plant in their area, reflecting a fear of something unknown and identified with a controversy.

As with space exploration, males, persons who had experienced a college science course, and graduate degree holders were all significantly more likely to be supportive of a local nuclear power plant than their counterparts; but all of these attributes were also positively associated with attentiveness to the nuclear power controversy in the first place. To identify the structure of these policy attitudes toward nuclear power, a stepwise logit analysis was performed. The dependent variable was dichotomized into those persons willing to accept a nuclear power plant in their area versus all others. The five independent variables were dichotomized as in the preceding logit models. The results indicated that gender was the most influential variable, with a significantly higher proportion of males favoring the placement of a local nuclear power plant than females. Holding constant exposure to college science courses and the level of formal education, gender accounted for a full 54 percent of the total mutual dependence in the model (see table 49). In contrast, attentiveness to science policy accounted for only 10 percent of the total mutual dependence and college science course exposure explained 9 percent. Attentiveness to the nuclear power issue itself was not significantly associated with the policy attitude when the other four independent variables were taken into account.

The major issue concerning chemical food additives has been the potential for harm to the health of consumers. If a person views this matter as one of little personal consequence, then it is unlikely that he or she would devote any significant time or effort to learning more about the issue or would take any steps to avoid any particular additives or foods. Each respondent was asked in the 1979 study to indicate whether chemical food additives were a serious problem, a not very serious problem, or not a problem at all. The results indicated that 29 percent of the public thought that chemical food additives were a serious problem, 52 percent viewed them as a minor problem, and 19 percent saw no problem at all (see table 51). In contrast, 44 percent of those persons attentive to science policy and food additives viewed the problem as serious, and 49

TABLE 51
POLICY PREFERENCE CONCERNING FOOD ADDITIVES: 1979

	Chemical Additives to Food Are . . .			
	Not a Problem	A Problem, but Not Serious	A Serious Problem	N
Total	19%	52%	29%	1635
Attentive to . . .				
Food additives & science	5	51	44	151
Food additives only	5	47	49	171
Science only	19	64	17	256
Neither area	25	51	25	1057
Age				
17–24	19	55	27	309
25–34	14	54	32	361
35–44	16	50	35	258
45–54	18	56	26	234
55–64	23	46	31	228
65 and over	27	48	24	245
Gender				
Female	16	50	34	862
Male	22	54	24	773
Education				
Less than high school	30	41	29	465
High school	16	56	29	932
Baccalaureate	13	53	34	146
Graduate degree	8	63	29	92
College Science Courses				
None	21	51	28	1210
Some	14	53	33	425

percent of those individuals attentive to the food additive issue alone classified the problem as serious. Only 17 percent of science policy attentives who did not follow the food additive controversy assessed the problem as serious. Approximately 25 percent of those persons who were not attentive to either science policy or the food additive controversy reported that they did not think that there was a problem at all.

Females and persons who had taken a college science course were slightly more likely to view the food additive issue as serious, but both of these factors were also associated with attentiveness to the food additive issue itself. To identify the relative contribution of each of these attributes, a stepwise logit analysis was performed. The dependent variable was dichotomized into those individuals who viewed the problem as serious versus all others. The five independent variables were dichotomized as in the preceding models. The results indicated that attentiveness to the food additive issue was the most influential variable, accounting for 52 percent of the total mutual dependence when all of the other independent variables were held constant (see table 49). Gender accounted for an additional 15 percent of the total mutual dependence. Neither college science course exposure, the completion of a baccalaureate, nor attentiveness to science policy were significantly associated with the individual's assessment of the seriousness of the food additive issue.

In summary, these analyses have indicated that specific issue attentiveness was significantly associated with both space exploration and food additive policy views but not with the nuclear power policy preference. In all three controversies, gender was a significant factor. Exposure to a college science course was a significant factor in the determination of policy attitudes toward space exploration but only a minor influence on policy views concerning the other two controversies. Once exposure to college science courses was taken into account, there was no significant residual effect from the completion of a baccalaureate for any of the three specific issue areas. Attentiveness to science policy appeared to be strongly associated with substantive policy views in regard to space exploration, weakly associated with nuclear power policy thinking, and unrelated to the food additive controversy.

THE RESOLUTION OF SPECIFIC ISSUE CONTROVERSIES

In the earlier discussion concerning the public attitudes toward science policy, the attentive public was portrayed as responsive to the discussions and initiatives of the nongovernmental leaders of science policy. It was suggested that the science policy agenda was set largely by the leadership group and that the views of the attentive public were influential to the extent that they contacted decision makers to express agreement or disagreement with specific policy positions advocated by leaders or policy makers.

In contrast to the more homogeneous views of the leadership group for science policy, the narrower issue controversies active in each of the three areas utilized in the preceding analysis are characterized by more polarized leaders who advocate significantly different policy outcomes. In this case, it is important to identify the policy-making groups that the members of these attentive publics view as competent and appropriate to resolve these disputes. Given a set of competing leaders in any specific controversy, it is likely that the attentive public for that specific issue would tend to focus on the views and perhaps the data provided by those groups that they view as appropriate issue judges. Following this logic, the 1979 survey of public attitudes toward science and technology posed a hypothetical conflict in each of the three specific issue areas utilized in the preceding analysis and asked each respondent to identify those individuals or groups that they judged to be the most qualified to resolve the issue.

In the area of space exploration, the survey posed a hypothetical conflict between "one group of scientists" who want to send radio messages into deep space to try to communicate with other civilizations and "another group of scientists" who believe that this could lead to the conquest of our own civilization by more advanced groups that might receive the signal. Each respondent was shown a card listing a wide range of groups that might be considered appropriate for resolving the dispute and was asked to identify the group that was the "most qualified" to make the decision. The respondent was also asked to identify the "next most-qualified" group and the "least-qualified" group for resolving the disagreement.

In the area of nuclear power, the 1979 study posed a classic siting dispute pitting "an electric utility company" that wanted to build a nuclear power plant in a particular town or county versus "a group of local citizens" who were afraid that it might be dangerous and organized to stop its construction. Each respondent was shown a card listing several groups and agencies that have been involved in previous siting conflicts and was asked to make the same selections outlined above in regard to space exploration.

In the area of food additives, a hypothetical dispute was described between "a food-processing company" that wanted to add a chemical to one of its products to make it more resistant to spoilage and a "federal agency" whose tests raised questions about the safety of the additive. Again, each respondent was given a card with a wide range of groups and agencies that have been involved in similar disputes in recent years, and the respondent was asked to select the most-, next most-, and least-qualified groups to make a final decision on the dispute.

In all three issue areas, the most frequently cited group was "scientists and engineers who specialize in this area," and this preference was expressed by both the attentive and nonattentive publics in all three specific issue areas (see table 52). After the selection of scientists and engineers knowledgeable in an area, the selection of other qualified groups differed significantly by the nature of the specific issue area.

In the space communication dispute, the "administrators of NASA" were

TABLE 52
BEST-QUALIFIED GROUPS TO RESOLVE SPECIFIC CONTROVERSIES: 1979

	Space Exploration		Nuclear Power		Food Additives	
	Attentive Public	Non-Attentive Public	Attentive Public	Non-Attentive Public	Attentive Public	Non-Attentive Public
A group of scientists and engineers who specialize in area	74%	69%	63%	58%	92%	83%
The food processing companies	NA	NA	NA	NA	16	28
NASA administrators	64	66	NA	NA	NA	NA
Utility company that will operate plant	NA	NA	11	23	NA	NA
Local government officials	1	3	16	15	1	5
The governor and state legislature	1	3	7	9	2	3
The president and Congress	14	15	3	6	1	7
A federal regulatory agency or commission	NA	NA	38	30	59	43
Citizens voting in a referendum	22	17	50	43	NA	NA
The United Nations	21	14	NA	NA	NA	NA
Consumers who buy that kind of product	NA	NA	NA	NA	28	29
N =	137	1498	423	1212	407	1228

Note: The responses in this table reflect the respondent's judgment of the "best-qualified" and the "next best-qualified" groups. The percentages may sum up to more than 100 percent.

the second most favored decision makers, and this preference was expressed by about two-thirds of both the attentive and nonattentive publics for space exploration (see table 52). No other group was preferred by as many as a quarter of the attentive or nonattentive publics, including the president, the Congress, the United Nations, or citizens voting in a referendum.

In the nuclear power plant siting dispute, the second most preferred group was "the citizens of the community voting in a referendum" (see table 52). The referendum choice was supported by 50 percent of the attentive public for nuclear power issues and 43 percent of the nonattentive public. The third most popular choice for resolving the siting issue was a "federal regulatory commission," which was favored by 38 percent of the attentive public for nuclear power issues and 30 percent of the nonattentive public for that issue.

In regard to the food additive dispute, the second most preferred group was a "federal regulatory agency," which was preferred by 59 percent of the attentive public for the food additive area and 43 percent of the nonattentive public for that issue (see table 52). Consumers were judged the third most competent group, but were favored by only 28 percent of the attentive public and 29 percent of the nonattentive public. There was little support for the food-processing company as a decision maker by either the attentive or nonattentive publics.

In summary, these results indicate that there was a high degree of consensus among the public on the desirability of utilizing scientific and technical advice on the resolution of specific science-oriented disputes. There was little fear of the federal government *per se*, with substantial support for a decision making role for NASA in the space dispute and for the appropriate federal regulatory agencies for both nuclear power and food additives. The idea that there is a strong public commitment to state and local government as decision makers wherever possible was not supported by these results, with very low levels of support expressed for a decision making role for state or local public officials. The private users of science and technology, the utility and food-processing companies, were not viewed as appropriate decision makers on these matters by substantial majorities of both the attentive and nonattentive public.

PERSONAL PARTICIPATION IN SPECIFIC CONTROVERSIES

In the context of the general model, some of the members of an attentive public would be expected to be sufficiently concerned about an issue or problem that they would take a personal role in the dispute. In most cases, this personal participation would be focused on contacting decision makers by letter or telephone to urge support for a given policy outcome. In other cases, participation might include contributing funds or time to an organization supporting a particular policy position, signing petitions, attending conferences and rallies, or participating in demonstrations and protests. The process of convincing members of the attentive public to engage in any of these activities is referred to as mobilization. Rosenau (1974) has provided an excellent description of the mobilization process in regard to both a civil rights issue and a foreign policy issue. No previous studies of mobilization of the attentive public for science policy or of attentive publics for narrower science-related controversies have been reported in the literature.

In recognition of this void, the 1979 survey of public attitudes toward science and technology was designed to measure the likelihood of personal participation in each of the three disputes discussed above. As a preface to the analysis of those data, however, it is necessary to examine the general level of social and political participation reported by each of these three attentive publics.

In terms of participation in voluntary organizations, the attentive publics for space exploration, nuclear power, and food additives displayed a relatively high level of activity (see table 53). It will be recalled that 73 percent of the attentive public for science policy reported an active role in one or more voluntary organizations (see table 23), and essentially the same level of activity was reported by each of these three specific issue attentive publics. The higher proportion of food additive attentives active in religious organizations and the lower proportion of food additive attentives active in labor unions and political

TABLE 53
SOCIAL AND POLITICAL BEHAVIOR OF SPECIFIC ISSUE ATTENTIVE PUBLICS: 1979

	Attentive Public for . . .		
	Space Exploration	Nuclear Power	Food Additives
Active member in . . .			
Church or religious group	38%	39%	46%
Professional or business association	35	30	26
Labor union	14	11	6
Agricultural association	3	5	4
Political organization	14	8	8
Local civic association	22	21	26
Service club or fraternal organization	29	25	18
Art, drama, or sports group	31	29	28
No voluntary organizations	24	29	28
Social Trust Index			
0 (low)	29	21	19
1	26	20	25
2	21	28	23
3 (high)	24	32	33
Contacted public official on policy-related matter	45	39	37
N =	137	423	407

organizations reflected, in part, the higher proportion of women in the attentive public for food additive issues.

The attentive publics for nuclear power and food additives displayed a relatively high level of social trust, with 60 and 56 percent, respectively, scoring in the upper half of the Index of Social Trust (see table 53). This is comparable to the 58 percent of the attentive public for science policy that scored in the upper half of the index. In contrast, only 45 percent of the attentive public for space exploration scored in the upper half of the index, a significantly lower proportion than either the attentive public for science policy or the other two specific issue attentive publics. The reason for this relatively lower level of social trust among space exploration attentives is not immediately apparent.

When the level of political activity of each of the specific issue attentive publics was viewed in the context of the political-specialization process discussed in chapter 2, all three attentive publics displayed a relatively high level of political activity, but slightly lower than that reported by the attentive public for science policy. Fifty-five percent of the attentive public for space exploration were single- or multi-issue activists, as were 52 percent of the nuclear power attentives and 48 percent of the food additive attentives (see table 54). Just over two-thirds of the attentive public for science policy were classified in those two groups in 1979. While the level of activist involvement was slightly lower among the three specific issue attentive publics than the attentive public for science policy, it is necessary to recall that in that same year the proportion of the public not attentive to science policy that was included in those two groups was only 18 percent (see table 24).

In terms of policy-related contacts with decision makers, the level of activity reported by the three specific issue attentive publics did not differ significantly among themselves (see table 53) or from the level reported by the attentive public for science policy in the same year. The rate of contacting reported

TABLE 54
LEVELS OF POLITICAL PARTICIPATION FOR SPECIFIC ISSUE ATTENTIVE PUBLICS: 1979

	Attentive Public for . . .		
	Space Exploration	Nuclear Power	Food Additives
Inactive citizens	11%	10%	14%
Ritual voters	7	8	9
Issue voters	13	15	16
Ritual actives	16	15	14
Single-issue actives	19	22	20
Multi-issue actives	35	30	28
N =	137	423	407

by each of the three specific issue attentive publics was at least twice as high as the contacting rate for persons not attentive to that specific issue.

Looking at the general level of political participation for the three specific issue attentive publics, the data from the 1979 study indicated: that these groups were relatively more active in voluntary associations than their nonattentive counterparts, that they were more likely to be political activists, and that they were more likely to have contacted a public official on a policy-related matter than persons not attentive to science policy issues. Given this general background, it is now appropriate to turn to an analysis of the likelihood of personal participation in each of the three controversies discussed in previous sections.

Turning first to the dispute over the transmission of radio signals into deep space, each respondent in the 1979 study was asked to assess the likelihood of their personal participation in that dispute and to indicate whether they would definitely, probably, probably not, or definitely not participate in the dispute. Only 7 percent of the total adult population indicated that they would definitely participate in the controversy, but 26 percent of those persons attentive to both space and science and 12 percent of those attentive to space exploration alone indicated that they would definitely participate in the matter (see table 55). Eleven percent of those persons attentive to science but not to space reported that they would definitely take a role in the dispute, but only 5 percent of those persons attentive to neither space nor science indicated a definite plan to participate. Males and persons with some exposure to a college science course were also more likely to take part in the controversy.

To better understand the relative contribution of these factors in the decision to participate in the space controversy, a stepwise logit analysis was conducted. The intention to participate was dichotomized into those who would definitely participate versus all others. The five independent variables were dichotomized according to the same procedures utilized in the preceding logit analyses. The results indicated that gender accounted for 25 percent of the total mutual dependence in the model and that attentiveness to science policy accounted for another 22 percent of the mutual dependence (see table 56). About 18 percent of the total mutual dependence was associated with exposure to a college science course, and 16 percent was related to attentiveness to space exploration itself. When exposure to a college science course was held constant, the logit model indicated that there was no residual explanatory power associated with the completion of a baccalaureate degree.

Some additional insight into the decision to participate personally in this dispute can be gained by examining the reasons given by the respondents for nonparticipation. In the 1979 interview, all respondents who indicated that they probably or definitely would not participate in the hypothetical dispute were given a card that listed several possible reasons for not participating, and the respondent was asked to indicate all of the reasons that were applicable in his or her case. Among those respondents not attentive to space exploration, 69 percent of the nonparticipants indicated that they did not know enough about

TABLE 55
PARTICIPATION IN A CONTROVERSY CONCERNING SPACE EXPLORATION: 1979

	Likelihood of Personal Participation				
	Definitely Would	Probably Would	Probably Not	Definitely Not	N
Total	7%	26%	43%	25%	1635
Attentive to . . .					
Space & science	26	52	20	3	79
Space only	12	42	42	4	58
Science only	11	39	41	10	243
Neither area	5	21	45	30	1255
Age					
17–24	8	30	45	18	309
25–34	11	29	44	16	361
35–44	7	26	49	18	258
45–54	7	27	44	21	234
55–64	5	24	44	28	228
65 and over	2	16	30	52	245
Gender					
Female	4	23	46	27	862
Male	10	29	40	21	773
Education					
Less than high school	7	19	31	43	465
High school	6	28	47	19	932
Baccalaureate	10	28	50	12	146
Graduate degree	6	30	52	12	92
College Science Courses					
None	5	23	43	29	1210
Some	11	33	44	13	425

the issue to participate (see table 57). Twenty-eight percent indicated that they did not think that participation would "do any good," and 21 percent said that they would not know whom to contact. In contrast, slightly over a third of the attentives who would have resisted mobilization on this issue indicated that participation would not be effective, they would not know whom to contact, they did not know enough about the issue, and that someone else would probably express their views. The difference in emphasis between these two sets of reasons is important. A substantial majority of the nonattentive nonparticipants stressed their feeling that they did not know enough about the issue itself and that they would not know whom to contact. Both substantively and structurally, they suffered from a deficit of information. In contrast, most attentives would have participated, and the attentives that would not have participated tended to emphasize cynicism about the efficacy of participation generally and of their personal involvement. These profiles fit the general model of issue attentiveness described in chapters 2 and 3.

In sharp contrast to the relatively low-key nature of the hypothetical space

TABLE 56
THREE LOGIT MODELS TO PREDICT PARTICIPATION IN SPECIFIC CONTROVERSIES: 1979

Models		Coefficient of Multiple-Partial Determination When Y Represents . . .		
		Space Exploration	Nuclear Power	Food Additives
H1	AZGEC,YC	.183	.293	.113
	Difference due to YC	.183	.293	.113
H2	AZGEC,YC,YE	.184	.344	.131
	Difference due to YE	.001	.050	.018
H3	AZGEC,YC,YE,YG	.434	.346	.398
	Difference due to YG	.250	.002	.267
H4	AZGEC,YC,YE,YG,YZ	.650	.456	.404
	Difference due to YZ	.215	.110	.006
H5	AZGEC,YC,YE,YG,YZ,YA	.807	.723	.868
	Difference due to YA	.157	.267	.464

Y = personal participation
C = college science courses
E = education
G = gender
Z = attentiveness to science policy
A = attentiveness to specific issue

communications controversy, there have been a number of actual controversies over the siting of nuclear power plants in recent years. In the 1979 study, each respondent was asked to consider a hypothetical conflict between an electric utility company that wanted to build a new nuclear power plant in a given locale and a local citizens group that expressed concern about the safety of the plant and opposed its construction. Following the same format utilized in the space communication controversy, each respondent was asked to report whether he or she would definitely, probably, probably not, or definitely not participate in a dispute like this in his or her own area. Almost a quarter of the total public indicated that they would definitely participate, and an additional 35 percent reported that they would probably take part in the dispute (see table 58). Just over 40 percent of those respondents who were attentive to both science policy and nuclear power indicated that they would definitely participate, while 35 percent of those persons attentive to only the nuclear power issue definitely would have taken a role in the controversy. This is a significantly higher level of direct involvement than was reported in the space communications controversy, suggesting that proximity of the issue may increase personal participation significantly. Although persons with higher levels of formal education and persons who had experienced a college science course were more likely to express a definite commitment to participate than their counterparts, the margin of dif-

TABLE 57
REASONS FOR NOT PARTICIPATING IN A CONTROVERSY OVER SPACE EXPLORATION: 1979

	Attentive Public for Space Exploration	Nonattentive Public for Space Exploration
It wouldn't do any good	35%	28%
I wouldn't know who to contact	37	21
Someone else would probably express my views	36	14
I don't know enough about the issue	36	69
I have too many other things to do	22	19
It would not affect me personally	4	14
N =	45	1057

Note: Respondents were allowed to select as many reasons as they wished, so percentages may sum up to more than 100 percent.

ference was not very large. There was no significant gender difference in the proportion of men and women who expressed a commitment to take a personal role in this type of dispute.

To better understand the structure of personal participation in the nuclear power siting dispute, a stepwise logit analysis was performed. The dependent variable was dichotomized into those persons who reported a definite intention to participate versus all others. The five independent variables were dichotomized as in the previous logit analyses. The results indicated that exposure to a college science course accounted for 29 percent of the total mutual dependence in the model and that attentiveness to the nuclear power issue itself explained an additional 27 percent of the mutual dependence (see table 56). Attentiveness to science policy accounted for 11 percent of the total mutual dependence, but neither gender nor the completion of a baccalaureate degree had a significant residual relationship with personal participation in the hypothetical siting dispute.

An examination of the reasons for nonparticipation revealed a pattern similar to that found in regard to the space communications issue. Sixty percent of

TABLE 58
PARTICIPATION IN A CONTROVERSY CONCERNING NUCLEAR POWER: 1979

	Likelihood of Personal Participation				
	Definitely Would	Probably Would	Probably Not	Definitely Not	N
Total	24%	35%	27%	15%	1635
Attentive to . . .					
Nuclear power & science	42	41	14	3	179
Nuclear power only	35	42	21	3	244
Science only	27	39	29	5	143
Neither area	18	32	30	20	1069
Age					
17–24	21	42	32	5	309
25–34	33	36	23	8	361
35–44	29	35	27	9	258
45–54	28	37	24	10	234
55–64	17	36	31	16	228
65 and over	12	24	24	41	245
Gender					
Female	23	35	28	15	862
Male	25	36	26	13	773
Education					
Less than high school	16	24	29	31	465
High school	25	40	27	8	932
Baccalaureate	40	33	23	4	146
Graduate degree	33	42	23	2	92
College Science Courses					
None	21	34	28	17	1210
Some	34	40	22	4	425

the public not attentive to the nuclear power issue indicated that they did not know enough about the issue to participate, and 15 percent reported that they would not know whom to contact on this issue (see table 59). In contrast, only 29 percent of those attentive to the nuclear power issue claimed inadequate knowledge to support personal participation. Among attentives there was a feeling that it would not be effective (34 percent), someone else would express their view (25 percent), and that they had too many other things to do (22 percent). The failure of these attentive individuals to respond to the mobilization stimulus on this issue appears to reflect the intense competition for the individual's time and a negative assessment of the probable efficacy of personal involvement in this issue. When there is a critical mass already active on any issue, it is easier for a busy person to conclude that his or her own involvement might have a relatively small marginal effect; whereas participation in a less popular issue, like space communications, might have a relatively larger marginal impact.

Looking at the overall profile of public participation in a hypothetical nu-

TABLE 59
REASONS FOR NOT PARTICIPATING IN A CONTROVERSY OVER NUCLEAR POWER: 1979

	Attentive Public for Nuclear Power	Nonattentive Public for Nuclear Power
It wouldn't do any good	34%	30%
I wouldn't know who to contact	11	15
Someone else would probably express my views	25	17
I don't know enough about the issue	29	60
I have too many other things to do	22	17
It would not affect me personally	12	9
N =	87	577

Note: Respondents were allowed to select as many reasons as they wished, so percentages may sum up to more than 100 percent.

clear siting dispute, the results of the 1979 study indicated that there would be a relatively high level of public involvement and that attentiveness to the nuclear power issue and prior exposure to a college science course were the two most important factors in the individual's decision about personal participation in the issue. An absence of substantive knowledge about the nuclear power issue appeared to be important for the nonattentive segment of the public, while the attentive segment of the public tended to doubt the efficacy and the marginal utility of their personal involvement.

The hypothetical dispute over the addition of a chemical preservative to a food product was selected because it reflected the opportunity for individuals to make a decision on the issue as consumers as well as politically as citizens. As noted above, substantial majorities of both the attentive and nonattentive publics indicated that they would prefer to have this type of dispute settled by scientists and engineers who specialize in the area, or secondly, by a federal regulatory agency. Although these segments of the public might have preferred to have the dispute settled by experts or regulators, most people are exposed to a wide array of news about scientific studies concerning the safety of various chemical additives to foods and beverages and must decide for themselves whether this information is sufficiently serious to merit a change in their personal shopping and eating habits.

To assess the extent to which consumers actually change their personal behaviors in reaction to news about the safety of various additives, the 1979 study asked each respondent if he or she had changed his or her own shopping or eating habits due to reports about possible dangers associated with a food additive. The results indicated that 46 percent of the total adult population indicated that they had made some change in their buying or eating habits for that reason (see table 60). Approximately 70 percent of those persons attentive to the food additive issue reported that they had changed their personal shopping or eating habits due to concern about the safety of an additive. Females, persons who completed a baccalaureate, and persons who had experienced at least one college science course were all somewhat more likely to have changed their personal behaviors than were their counterparts.

To understand the relative influence of these factors in the decision to

TABLE 60
CHANGE IN SHOPPING OR EATING HABITS OVER FOOD ADDITIVES ISSUE: 1979

	Percentage Changed Shopping or Eating Habits	N
Total	46%	1635
Attentive to . . .		
Food additives & science	68	151
Food additives only	70	171
Science only	34	256
Neither area	39	1057
Age		
17–24	39	309
25–34	53	361
35–44	50	258
45–54	42	234
55–64	49	228
65 and over	39	245
Gender		
Female	53	863
Male	37	773
Education		
Less than high school	40	465
High school	45	932
Baccalaureate	60	146
Graduate degree	54	92
College Science Courses		
None	42	1210
Some	56	425

change one's shopping or eating habits, a stepwise logit analysis was performed. The dependent variable was dichotomized into those persons who reported a change in their shopping or eating habits versus those who made no change. The five independent variables were dichotomized as in the preceding logit analyses. The results indicated that attentiveness to the food additive controversy itself was the most important factor in the model, accounting for 45 percent of the total mutual dependence (see table 56). Gender accounted for an additional 27 percent of the total mutual dependence in the model, while exposure to a college science course explained another 11 percent of the mutual dependence. The completion of a baccalaureate or attentiveness to science policy had no residual association with changes in shopping and eating habits when the other independent variables in the model were held constant.

Looking at the patterns of probable participation in the food additive controversy, it appears that almost half of the adult population have actually modified their shopping or eating habits in response to reports about the potential dangers of some chemical food additives. While the dispute was presented in a general form and would not necessarily apply to any given food additive controversy, the results do suggest that a significant proportion of the population is sensitive to this issue and willing to make changes in their personal habits. The logit analysis indicated that attentiveness to the food additive issue itself is the best predictor of the likelihood of making a change in shopping or eating behaviors. Gender was the second best predictor, with a significantly higher proportion of women reporting a change in their shopping or eating habits than men.

SUMMARY

This analysis has shown that there are identifiable attentive publics that focus on specific controversies within the broader domain of science policy but that do not necessarily follow science policy itself. Similarly, the results indicated that persons who were attentive to science policy generally were not necessarily attentive to every narrower controversy that is related to science and technology. Specific attentive publics were identified for space exploration, nuclear power, and food additives. Generally, these specific issue attentive publics resembled the attentive public for science policy, but there were some notable exceptions. In the food additive dispute, women were as likely to be attentive to this issue as men; whereas women had been significantly less likely than men to be attentive to science policy, nuclear power, or space exploration.

In their substantive policy views, these specific issue attentive publics: favored the continuation of the space program, displayed substantial reservations about the siting of a nuclear power plant in their area, and were more likely to view the food additive issue as a serious problem than persons not attentive to that issue. In all cases, the majority of the attentive and nonattentive publics held common policy views, but a higher proportion of the attentive public held that view than the corresponding nonattentive public.

There was also a high degree of consensus between the attentive and nonattentive publics in all three specific controversies concerning the individuals or groups best qualified to resolve those disputes. In all three areas, the majority choice of both publics was "scientists and engineers who specialize in that area." In the nuclear power siting dispute, a citizen referendum was the second most popular method for resolving the dispute, but in the other two areas federal agencies were given the second highest rating. The idea that people prefer local officials to resolve local disputes did not appear to apply to these specific controversies.

In a pattern similar to the attentive public for science policy, all three specific issue attentive publics displayed a relatively high level of political activity. The members of these specific issue attentive publics were usually members of one or more voluntary organizations, reflecting a higher than average level of social trust. A significant portion of all three specific issue attentive publics were active in contacting public officials on policy-related matters.

In terms of direct participation in the specific controversies, the analysis indicated that attentiveness is a major factor in the successful mobilization of citizens. In a relatively low-salience area like the space communication dispute, approximately 20 percent of the attentive public for space exploration expressed a definite intention to participate in contrast to only 5 percent of those persons not attentive to science or space. A similar, but higher, pattern was found in regard to the nuclear power issue. In the food additive controversy, attentiveness to that issue was the major predictor of participation. Among attentives, the major reasons for resisting mobilization on these issues appeared to involve an intense competition for the individual's time and a negative assessment of the probable efficacy of personal involvement.

While these three controversies were more concrete and offered more direct opportunities for participation than many more general science policy issues, the general patterns of participation observed here should hold for the personal participation of the attentive public for science policy. The differences observed in the patterns of attentiveness, policy views, and probable participation in these three specific controversies illustrate the diversity of paths to attentiveness and mobilization.

Chapter 9
The Future of Public Participation in Science Policy

Over the last two decades, an important social science literature has developed concerning the formation of attitudes during childhood and young adulthood and the persistence of some of these general dispositions and political attitudes into adulthood (Hess and Torney, 1967; Easton and Dennis, 1969; Dawson, Prewitt, and Dawson, 1978; Jennings and Niemi, 1974, 1981). Most often referred to as political socialization, these studies have indicated that basic orientations toward the political system are formed relatively early in life. While a full review of that literature is beyond the scope of this chapter, it is important to recognize that a number of important life goals are set during the young adult years and that these goals may have a significant impact on the relative interest accorded political activities generally (the process referred to as interest specialization) and on the specific issues that are perceived to be important (the process referred to as issue specialization). The decisions made during the young adult years are not irreversible, but the literature suggests that those patterns that are rooted in important life goals are more likelyto persist than choices that are less central to an individual's values (Allport, 1954; Hyman, 1959; Hyman, Wright, and Reed, 1975).

This chapter will utilize data from the 1978 National Public Affairs Study (NPAS) to explore the degree of the development of issue attentiveness in general and of attentiveness to organized science matters in particular. The 1978 study was designed to examine the development of issue specialization during the young adult years and included extensive batteries of items concerning foreign policy, science and technology, economic policy, and civil rights issues. For a more complete discussion of the study and the items measured, see the Appendix and Miller, Suchner, and Voelker (1980).

In the following analysis, the term "organized science" will be used to describe the object of attentiveness among young adults rather than the science policy label used in the adult analyses. This differentiation of terms is made to emphasize that the focus of young adult attentiveness is not as finely honed as adult attentiveness and that it is not reasonable to expect teenagers to be knowledgeable about the reasons for, and probable outcomes of, specific science policy matters. For those young adults interested in, and knowledgeable about, science and technology generally, the focus of their attentiveness appears to be a mixture of academic science derived from formal course experiences and some recognition of the potential uses or applications of scientific knowledge. If the young adult perceives this activity to be important and is willing to devote a portion of his or her time to becoming informed about it, that should be sufficient to indicate that the individual has a higher than average probability of following scientific issues in later years, especially when buttressed by career objectives or life goals. For high school students especially, it would be unreasonable to expect a thorough understanding of the structure of the scientific community or the dynamics of the scientific process. Accordingly, the term ''organized science'' is meant to include both basic and applied science and that broad array of scientific activities engaged in by the scientific community.

By seeking to understand the structure of young adult attitudes toward organized science, it will be possible to project the general contours of public attitudes toward science policy in the future. While the fit will not be perfect and numerous influential events will undoubtedly occur in the intervening years, it is likely that the basic life goals formed during the young adult years will influence a number of other important personal and career choices in the years immediately ahead, and that these decisions will in turn influence the level of interest in political affairs *per se* as well as the range of issues that will be viewed as important by a substantial segment of this young adult cohort.

THE EMERGENCE OF ATTENTIVENESS

During the high school and college years, young adults are expected to make a number of basic and far-reaching decisions about the level of education they will complete, the occupation or profession they will pursue, the area of the country in which they would prefer to live, the selection of a mate, and the initiation of a family. During this same period, these young adults are learning about the politics of their society and making some preliminary judgments about whether the results of political involvement are worth the investment of time and resources necessary to participate effectively. For most of the postwar decades, survey data have indicated that a significant proportion of the young adult population has not made a major commitment to political involvement, although a life-cycle interpretation of these data would suggest that they will become active when they have finished their schooling and have become tax-paying members of the work force. Without seeking to resolve all of the outstanding disputes in the literature on these matters, it is apparent that the young adult years are a critical period during which numerous important decisions are reached.

For the purpose of this analysis, the central issue is whether there is any evidence that the selection of areas of issue interest emerges during this period, and, if so, what are the dimensions and direction of these early issue attitudes concerning organized science. To answer these questions, it will be necessary to first examine the degree of issue specialization found in these young adults.

The 1978 National Public Affairs Study was designed to examine this set of hypotheses. The questionnaire presented each respondent with a set of 32 hypothetical "headlines" like those that might be found in a magazine or newspaper and asked each respondent to indicate whether he or she would have definitely, probably, probably not, or definitely not read each headline and its accompanying story. A set of factor analyses indicated that there was a high degree of issue specialization among young adults, with the most specialized interest structure being displayed by college students. A brief review of the factor structure among college students may be helpful in understanding the nature and extent of the issue specialization process at this relatively early point in the life cycle.

Seven significant factors emerged from the analysis.[14] The first factor included five headlines concerning foreign policy, an indication of the specialized nature of that topic (see table 61). The second factor included five headlines that concerned missiles, space flights, new weapons, and UFOs, and was labeled a technology factor. The third factor included headlines concerning fashions, cancer, birth control, the nutritional value of breakfast cereals, and the impact of unemployment on the family. These headlines all concern female and family interests and the data indicated that young women were significantly more likely to indicate a high probability of reading these items than young men. The fourth factor included two sports headlines and young men in the study were significantly more likely to read these than were young women.

The fifth and sixth factors involved a mixture of domestic political and policy items. The fifth factor included items about energy, pollution, and inflation; while the sixth factor included several civil rights, affirmative action, job, and welfare items. While there are several interesting differentiations between the two sets of issues, it appears that the items on the fifth factor involved people in a more direct and personal manner, while the items on the sixth factor were somewhat more distant and less personal.

The seventh factor included three items concerning scientific research. This factor was labeled a science factor.

From this limited review of the factor structure of the issue interests of college students, it is apparent that there is a high degree of specialization at this point and that the specialization clusters are plausible and understandable.

[14] The relationships among the headlines were measured with the ordinal correlation coefficient gamma and the resulting matrix was factor analyzed using the SPSS PA2 factor program. Both orthogonal and oblique rotations were performed, and the results reported reflect the oblique rotation. For a general discussion of factor analysis and the selection of rotations, see Rummel (1972).

TABLE 61
THE STRUCTURE OF ISSUE INTEREST AMONG COLLEGE STUDENTS: 1978

	Oblique Factor Patterns						
Headlines	1	2	3	4	5	6	h^2
Foreign trade policy	.84	.05	.08	.03	−09	−01	.70
Conflict in southern Africa	.81	−.07	−.08	−.05	.08	.06	.64
Italian elections	.70	−.17	.05	−.02	−.02	.02	.59
New trade treaties	.65	−.06	−.00	−.02	.04	−.15	.63
Middle East peace	.59	−.21	−.01	−.03	.07	−.02	.55
Affirmative action	.54	.18	−.05	.01	.21	−.13	.55
Need for welfare reform	.46	.17	−.06	−.04	.26	−.20	.55
New long-range missile	.10	−.75	−.01	.00	.09	−.25	.69
Soviets launch space station	.07	−.59	.29	−.05	.15	−.06	.61
New weapons system	.16	−.57	.12	.16	.04	−.09	.62
Need for manned flight	.20	−.48	.42	.13	−.08	−.01	.63
Human cell modification research	.12	−.10	.50	−.16	.02	−.20	.52
Basic science research	.25	−.07	.50	−.03	−.11	−.26	.64
Cancer therapy drug	.07	.00	.46	−.03	.31	−.14	.64
Need for research funding	.46	.02	.46	.04	−.07	−.12	.68
Study of UFO sightings	−.16	−.28	.42	.12	.21	.03	.41
Strength of NFL	−.06	.02	.02	.90	−.02	−.00	.58
Baseball favorites	.03	.05	.01	.78	−.00	−.01	.56
Rules on sex discrimination	−.04	−.01	−.04	−.08	.66	−.20	.53
New birth control device	.02	.07	.30	−.12	.56	.08	.47
Freedom-of-speech case	.30	−.04	−.00	−.03	.55	−.02	.55
Police-search procedures	.21	−.16	−.01	.07	.51	.01	.44
Impact of unemployment	.18	.26	−.04	−.00	.38	−.29	.51
Create new jobs	.18	.07	−.13	.11	.38	−.38	.51
Energy crisis	−.05	−.13	−.17	.05	.06	−.82	.57
Chemical pollution	.13	.04	.32	−.11	−.09	−.59	.66
Pollution in Great Lakes	.02	.01	.24	−.08	.01	−.58	.55
Solar energy	.07	−.16	.30	−.05	−.06	−.55	.60
Policy to fight inflation	.23	.04	−.07	.01	.21	−.50	.59
New report on oil reserves	.23	−.16	.17	.06	−.08	−.46	.59
Preview of summer fashions	−.06	.37	.14	−.06	.34	.09	.35
Value of breakfast cereals	−.07	.29	.33	.03	.27	−.15	.36
Percent total variance	32.1	10.6	6.7	5.7	4.3	3.2	62.7
Percent common variance	56.3	16.8	9.5	8.4	5.4	3.6	

The issue interest structures of high school students were similar, regardless of college plans. Separate analyses of high school students planning to attend college and not expecting to go to college each produced six factors, and both analyses tended to combine science and foreign policy issues into the first factor. The high school factor analyses explained slightly less of the total variance in the matrix than did the college student analysis, indicating that the degree of specialization was somewhat lower for those groups. In general, however, the factor structures for the three groups had a high level of similarity and all three indicated that the issue specialization process was active at these relatively early points in the life cycle.

To measure attentiveness to organized science among young adults, separate measures of interest in science and technology issues, knowledge about those issues, and information acquisition relevant to science and technology issues were developed.

Interest was measured by a combination of direct inquiry about interest in science and technology issues and the reported headline reading preferences of each respondent.[15] To create a single measure of interest in organized science, respondents who qualified as interested in either science or technology issues were classified as interested in organized science. Using this procedure, the data from the 1978 NPAS indicated that approximately half of the high school students not planning to attend college were interested in organized science, about two-thirds of college-bound high school students were interested in organized science, and that 77 percent of college students were also interested in organized science (see table 62). While these estimates may appear to be high, it is important to recall that these young adults have grown up in a more scientific and technological culture than perhaps any previous generation of young adults on this planet.

The second dimension of attentiveness to organized science is the level of knowledge of the young adult about science and technology. To measure the

[15] The structure of headline-reading preferences from the preceding factor analysis was used to develop a three-item science interest scale and a four-item technology interest scale. To be classified as interested in science, a respondent had to report a high level of interest in "issues involving science and technology" and to indicate an intention to read one or more of the three science headlines. At the same time, it was recognized that the measure of direct interest involved only a trichotomy and that some respondents may have reported that they were "somewhat" interested in science and technology as a reaction to the "very interested" label of the top category. Accordingly, those young adults who indicated that they were somewhat interested in organized science issues and who would have read two of the three science headlines were also classified as having a high level of interest in science.

Interest in technology was measured in a similar manner. Young adults who reported that they were very interested in issues involving science and technology and who reported an intention to read one or more of the four technology headlines used in the scale were classified as interested in technology. Similarly, respondents who reported that they were somewhat interested in organized science issues and who would have read two or more of the technology headlines were also classified as interested in technology.

TABLE 62
THE MEASUREMENT OF ATTENTIVENESS AMONG YOUNG ADULTS: 1978

	High School		
	No Plans for College	College Bound	College Student
Interested in science or technology issues	48.6%	67.1%	77.1%
Minus young adults not knowledgeable	45.6	46.4	22.6
Minus young adults not regular information consumers	0.7	3.6	7.9
Attentive public for science policy	2.3	17.1	46.6
Interested public for science policy	46.4	50.0	30.6
N =	1218	1098	1422

level of knowledge, the 1978 NPAS included a short battery of items concerning basic scientific constructs and a short series of items concerning one applied science controversy: nuclear power.[16] To qualify as minimally informed about science and technology for the determination of attentiveness, it was necessary for a young adult to have correctly defined at least two of the basic constructs and to have provided at least two arguments, pro or con, concerning nuclear power. Only 32 percent of the total group of high school and college students could meet this criterion, but the distribution strongly favored college students. Virtually all of the non-college-bound high school students who had qualified as interested in science and technology issues failed to meet the knowledge criterion (see table 62). Almost two-thirds of the college-bound high school students who had qualified as interested in science and technology issues also failed to meet this criterion. About 30 percent of the college students who were interested in organized science failed to qualify as substantively knowledgeable.

[16] The 1978 study asked each respondent to provide a brief definition of a molecule, an organic chemical, an amoeba, and DNA. The questions were open-ended and the respondent was required to write a one or two line explanation of each item, that was subsequently coded as correct or incorrect by a coding team trained for this purpose. A measure of applied issue knowledge was obtained in the 1978 study by asking each respondent to list all of the arguments that he or she had heard (up to four) concerning the advantages of building more nuclear power plants and all of the arguments (up to four) heard against more nuclear power plants.

In view of the large proportion of interested young adults who failed to meet the knowledge criterion, it is important to discuss briefly this result and its meaning. With a result of this type, it would be reasonable to ask if the knowledge criterion was too difficult. A review of the coding requirements for the basic constructs, which were primarily responsible for the low scores, revealed genuinely lenient definitions of correctness. The criterion would appear to be a minimal level of competence for reading or understanding a large share of the material available concerning contemporary science and technology issues in the United States. The alternative conclusion must be that high school science training did not provide this basic construct knowledge to a substantial portion of this high school population.

The third dimension of the attentiveness typology requires some evidence of a regular pattern of relevant information acquisition. For this purpose, it was necessary for a young adult to report that he or she either read a newspaper (including the national news section) most of the time, watched a television newscast most of the time, read a weekly news magazine regularly, or read a science news magazine regularly. Any one information source would have met the minimal acquisition criterion. A small proportion of those young adults who had demonstrated a high level of interest in organized science issues and a minimal level of substantive knowledge were dropped from the attentiveness group for this reason, but the impact of the acquisition criterion was the least influential of the three dimensions (see table 62).

To qualify as attentive to organized science, a young adult had to demonstrate a high level of interest, a minimal level of substantive knowledge, and some evidence of relevant information acquisition. The data from the 1978 study indicated that only 2 percent of non-college-bound high school students met this set of requirements, that about 17 percent of college-bound high school students met the standard, and that 47 percent of college students qualified under the definition (see table 62). These results suggest that the family, school, and other factors associated with the decision to attend college were positively associated with attentiveness to organized science; but the college experience itself appeared to be the strongest factor in the development of attentiveness to organized science.

Following the same analytic framework utilized with the adult studies reported in earlier chapters, it is useful to identify an "interested public for organized science." For this purpose, all of those young adults who expressed a high level of interest in organized science but who failed to meet either the knowledge or acquisition criteria were classified as an "interested public" for organized science. This group included 46 percent of non-college-bound high school students, half of college-bound high school students, and about 30 percent of college students (see table 62).

The developmental sequence of attentiveness to organized science can be seen by examining the percentage of young adults who qualified as attentive at each grade level within the three educational cohorts. Among non-college-bound high school students, the proportion of students attentive to organized science was extremely low, and the slight increase from 1.5 percent in the 10th grade to 3.1 percent in the 12th grade was not statistically significant at the .05 level (see fig. 4). For analytic purposes, the level of attentiveness among non-college-bound high school students should be viewed as essentially unchanged during the high school years. For college-bound high school students, the proportion attentive to organized science was approximately 18 percent, and there was no significant pattern of change during the high school years.

Among college students, the proportion attentive to organized science was significantly higher. In all four classes, the proportion attentive was slightly under 50 percent, but the differences between the classes were not significant at the .05 level, indicating that no significant pattern of change occurred during the four college years (see fig. 4).

From these data, it appears that both the intention to attend college and the college experience itself contribute to the likelihood of being attentive to organized science. The greater likelihood that college-bound high school students will be attentive to organized science undoubtedly reflects a complex set of family, peer, school, and other variables that are all associated with the intention to attend college. A thorough analysis of the relative contribution of these factors

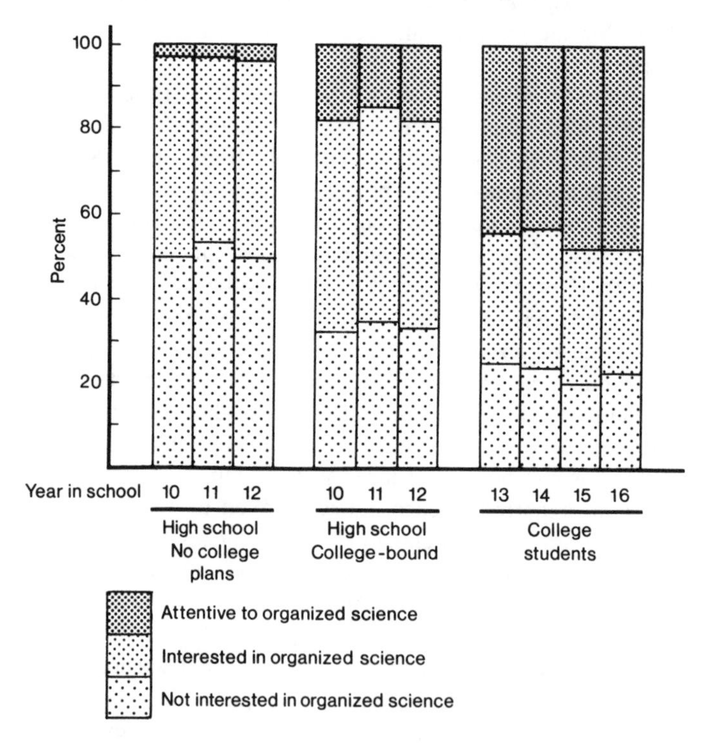

Figure 4: The Development of Attentiveness to Organized Science

to attentiveness has been provided by Miller, Suchner, and Voelker (1980). College attendance itself, however, appeared to provide a significantly larger stimulus to attentiveness to organized science. Recognizing the stable character of the level of attentiveness within each of the three school groupings, it would appear that college students were about three times more likely to be attentive to organized science than college-bound high school students. Miller, Suchner, and Voelker (1980) examined the family and background characteristics of the college-bound high school group and the college group and concluded that they reflected the same socio-economic segments of society and that there was no evidence of a substantial drop-out or change that would account for this significant differential. It appears that the differential must be attributed to the nature of the college experience itself. While a full analysis of this differential is beyond the scope of this chapter, it may be useful to speculate that the broad exposure of college undergraduates to science courses during the first two years and the greater political awareness and activity of most college campuses may contribute to this result.

In recognition of these patterns, a single variable has been created that reflects both the intention to attend college and actual attendance. Called educational status, this variable follows the procedure utilized above and divides respondents into three groups: (1) non-college-bound high school students, (2) college-bound high school students, and (3) college students. This variable will be used in most of the young adult analyses in this chapter.

An analysis of the demographic characteristics of the attentive group within each of the three educational status groups indicated that attentives were predominantly male, likely to be in the upper quarter of their class, and likely to have completed two or more science courses (see table 63). The finding that 6 percent of high school attentives and 14 percent of college attentives had not completed a science course at that level indicates that the attentiveness typology is not closed to persons without formal science training, but that formal science course exposure was positively and significantly associated with attentiveness.

To determine the relative contribution of each of these variables to attentiveness to organized science, a stepwise logit analysis was performed. The dependent variable was attentiveness, dichotomized into attentives and nonattentives. Science course exposure was a calculated variable that determined the typical number of science courses completed at each year of high school and college and then dichotomized the respondents into those who had completed an above-average number of science courses and those with an average or lower science course exposure.[17] Class rank was dichotomized into students who per-

[17] Since the modal number of science courses completed by students at each year of schooling would reflect the number of years in school, an adjustment was needed to take time into account. For this purpose, students were considered to have had an above-average science course exposure if they had completed two courses in grade 10, three courses in grade 11, four courses in grade 12, two courses as a college freshman, three courses as a college sophomore, or four courses as a college junior or senior.

TABLE 63
DEMOGRAPHIC PROFILE OF YOUNG ADULT ATTENTIVE PUBLIC FOR SCIENCE POLICY: 1978

	High School		
	No Plans for College	College Bound	College Student
Gender			
Female	28	32	35
Male	72	68	65
Class Rank			
Lower three quarters	47	19	41
Upper quarter	53	82	59
Science Courses			
None	6	6	14
One year	21	15	18
Two years	57	30	31
Three or more years	16	49	37
N =	28	187	662

TABLE 64
A LOGIT MODEL TO PREDICT YOUNG ADULT ATTENTIVENESS TO ORGANIZED SCIENCE: 1978

Models		df	LRX^2	CMPD
H1	RGSC,Y	23	1026.90	—
H2	RGSC,YC	22	964.66	.06
	Difference due to YC	1	62.24	.06
H3	RGSC,YC,YR	21	923.62	.10
	Difference due to YR	1	41.04	.04
H4	RGSC,YC,YR,YS	19	133.56	.87
	Difference due to YS	2	790.06	.77
H5	RGSC,YC,YR,YS,YG	18	18.37	.98
	Difference due to YG	1	115.19	.11

Y = attentiveness
C = science courses
R = perceived class rank
S = educational plans
G = gender

df = degrees of freedom
LRX^2 = likelihood ratio chi-square
CMPD = coefficient of multiple-partial determination

ceived themselves to be in the top 10 percent versus all others. Gender was dichotomized into male and female. Educational status was trichotomized as described above.

The stepwise logit analysis indicated that over three-quarters of the total mutual dependence in the model was accounted for by the educational status variable, holding constant science course exposure and perceived class rank (see table 64). The only other independent variable that accounted for a significant portion of the total mutual dependence was gender, which accounted for approximately 11 percent of the total mutual dependence. Class rank explained only 4 percent.

Science course exposure showed a weak relationship to attentiveness, even when entered into the stepwise model first. In view of the relatively important role that college science course exposure played in the stepwise logit models that predicted attentiveness to science policy among adults, this was a surprising result. One explanation may be that this measure included both high school and college science courses, whereas the previous measure for the adult analyses included only college science courses. In view of the questions raised earlier about the impact of high school science courses, this explanation may be particularly important.

In summary, these analyses have demonstrated that interest and issue specialization begin at relatively early points in the life cycle and that many high school and most college students displayed relatively clear specialized interest structures. This specialization process is reflected in the level of interest in, and knowledge about, organized science.

The data from the 1978 National Public Affairs Study indicated that relatively few non-college-bound high school students were attentive to organized science, although about 40 percent of this group reported a relatively high level of interest in science and technology topics. This discrepancy between a relatively high level of interest in organized science and a relatively low level of substantive science knowledge raises serious questions about the quality of high school science education for students not planning to attend college.

Approximately 18 percent of high school students planning to go on to college qualified as attentive, but when contrasted with the 44 percent of college freshmen who were attentive to organized science, this level also raises questions about the high school science experience. While the high school and college data were cross-sectional and not longitudinal in character, Miller, Suchner, and Voelker (1980) concluded that there were no significant differences in the socioeconomic or other characteristics of the two groups and that sampling or measurement problems could not be used to explain this discrepancy. It is important to note that the deficiency was not in interest about science and technology but rather in low levels of substantive knowledge about basic scientific constructs. An examination of the quality of the high school science experience would appear to be in order.

THE DEVELOPMENT OF ATTITUDES TOWARD ORGANIZED SCIENCE

Working within the attentiveness structure described in the preceding section, it is important to examine the substantive attitudes of young adults about organized science. It will be useful to divide this examination into an analysis of broad general attitudes, or dispositions, toward organized science and an analysis of student views on specific controversies active in 1978.

Looking first at general dispositions, the 1978 study indicated that young adults who were attentive to organized science were more likely to hold optimistic views about the benefits of science and scientific invention than nonattentive students (see table 65). This relationship held within all three of the educational status groups. A positive view of the benefits of organized science was held by a smaller proportion of college attentives than college-bound high school attentives. Two explanations are plausible. First, the college experience

TABLE 65
YOUNG ADULT DISPOSITIONS TOWARD SCIENCE AND TECHNOLOGY: 1978

	Attentive			Nonattentive		
Percentage Agreeing That . . .	HSNC	HSCB	COLL	HSNC	HSCB	COLL
Science is making our lives healthier, easier, and more comfortable	84%	86%	75%	57%	67%	69%
Scientific invention is largely responsible for our standard of living in the United States	92	99	94	82	93	93
One of the bad effects of science is that it breaks down people's ideas of right and wrong	43	24	11	41	32	15
One trouble with science is that it makes our lives change too fast	27	26	29	40	35	31
The growth of science means that a few people could control our lives	41	25	24	30	27	27
One of our big troubles is that we depend too much on science and not enough on faith	32	21	19	44	34	19
Overall, science and technology have caused more good than harm	76	77	71	54	64	64
N =	28	187	662	1190	910	760

HSNC = high school student, no college plans
HSCB = high school student, college-bound
COLL = college student

may have had a deflating effect on very positive views of the benefits of science and technology held by college-bound high school students. Second, since the proportion of attentives in the college group was significantly higher than the proportion for college-bound high school students, it may be that some of the college attentives were nonattentive in high school and still held some of the less positive dispositions associated with the nonattentive group. These explanations are not mutually exclusive.

In regard to potential negative effects from organized science, the 1978 data indicated that among college-bound high school students, those attentive to organized science were less likely to be concerned about potential negative effects than nonattentive students (see table 65). Only a small proportion of college students were concerned about potential negative effects, and this pattern did not vary by attentiveness. If one were to generalize from these cross-sectional data, it would appear that one effect of college is to reduce the proportion of students concerned about negative effects from science. Since the college experience also appeared to reduce the proportion of attentive students holding positive views of the benefits of science, the combination of these results would suggest that the college experience may result in a more balanced view of organized science.

To test this hypothesis, the 1978 study asked each student to evaluate the relative good and bad effects produced by science and technology. A significantly higher proportion of students attentive to organized science concluded that science and technology had produced more good than harm than did non-attentive students (see table 65). This relationship was significant within all three educational status groups. Among nonattentive students, those high school students expecting to go to college were generally more positive toward the results of science and technology than were non-college-bound high school students. Within attentiveness groups, however, there was no significant difference between college-bound high school students and college students, suggesting the absence of a college effect.

To better understand the factors associated with this view, a stepwise logit analysis was performed. The dependent variable was each student's agreement or disagreement with the statement that science and technology had caused more good than harm, which was treated as a dichotomous variable. Science course exposure, perceived rank in class, educational status, gender, and attentiveness to organized science were entered into the stepwise logit model in that order; and all were defined the same as in the previous logit model predicting attentiveness.

The results of the logit analysis indicated that educational status was the most important correlate of young adult attitudes toward the benefits and harms of science and technology, accounting for approximately 30 percent of the total mutual dependence in the model (see table 66). Attentiveness to organized science was the second most influential factor, explaining 12 percent of the total

TABLE 66
A LOGIT MODEL TO PREDICT YOUNG ADULT ASSESSMENT OF THE BENEFITS AND HARMS OF SCIENCE: 1978

Models		df	LRX^2	CMPD
H1	AGERC,Y	47	128.76	—
H2	AGERC,YC	46	126.09	.02
	Difference due to YC	1	2.67	.02
H3	AGERC,YC,YR	45	115.84	.10
	Difference due to YR	1	10.25	.08
H4	AGERC,YC,YR,YE	43	77.71	.40
	Difference due to YE	2	38.13	.30
H5	AGERC,YC,YR,YE,YG	42	66.49	.48
	Difference due to YG	1	11.22	.09
H6	AGERC,YC,YR,YE,YG,YA	41	51.52	.60
	Difference due to YA	1	14.97	.12

Y = agree more good than harm
C = science courses
R = perceived class rank
E = educational status
G = gender
A = attentiveness to science

df = degrees of freedom
LRX^2 = likelihood ratio chi-square
CMPD = coefficient of multiple-partial determination

mutual dependence with all of the other independent variables held constant. Gender accounted for 9 percent of the total mutual dependence, and perceived class rank accounted for an additional 8 percent of the total mutual dependence. Science course exposure explained only 2 percent of the total mutual dependence.

Turning to young adult views toward specific controversies, the 1978 study found that young adults were generally supportive of funding for scientific research, optimistic about the ability of science to solve the energy problem, concerned about the safety of nuclear power, and supportive of the space program (see table 67). A more detailed review of these specific attitudes may be helpful in understanding the structure and direction of these young adult attitudes.

As noted in the earlier discussion of the science policy agenda, adequate funding for basic scientific research was the highest priority among science policy leaders. The data from the 1978 study indicated that a majority of young adults felt that the federal government should "spend more money on scientific research." This view was held by a significantly higher proportion of attentives than nonattentives in all three educational status groups (see table 67). Approximately three-quarters of college students attentive to organized science held this view.

A majority of high school and college students expressed optimism that science and technology would find a long-term solution to the energy problem.

TABLE 67
YOUNG ADULT ATTITUDES ON SELECTED SCIENCE POLICY ISSUES: 1978

	Attentive			Non-Attentive		
Percentage Agreeing That . . .	HSNC	HSCB	COLL	HSNC	HSCB	COLL
The federal government should spend more money on science research	56%	74%	73%	45%	55%	50%
We can depend on science and technology for a long-term solution to the energy problem	62	67	65	47	55	54
Solar energy is the best single long-term solution to our energy problem	54	65	55	60	63	57
There are still major unanswered questions about the safety of nuclear power plants	87	87	82	81	87	84
The risk involved in generating nuclear power is relatively minor and should not block the construction of new nuclear power plants	27	37	27	28	25	19
A privately-owned chemical company should be allowed to manufacture and sell anything they want without government interference	9	3	4	15	8	4
When all is said and done, the space program still hasn't produced much of value for the U.S.	38	32	19	44	42	34
The U.S. has lost its lead in science research to the Soviet Union and other nations	21	21	11	22	21	13
It is important that the U.S. continue to develop new weapons at least as fast as the Russians	76	66	53	62	67	45
N =	28	187	662	1190	910	760

Again, a higher proportion of attentives held this view in all three educational status groups (see table 67). A slight majority of young adults also expressed the view that solar energy was the best long-term solution to the energy problem. There was no significant difference in the proportion of attentives and nonattentives holding positive views of solar power.

Approximately four of five young adults expressed concern about the safety of nuclear power plants. This pattern persisted across educational plan groupings and there was no significant difference between attentives and nonattentives on this issue (see table 67). Although the proportion of students expressing concern about nuclear power plant safety was uniform, a significantly higher proportion of college-bound high school and college attentives were willing to allow new nuclear power plant construction than nonattentives in the same educational status groups. For example, 27 percent of college attentives were willing to allow new construction in contrast to only 19 percent of college students not attentive to organized science. The reservations about the safety of nuclear power were consistent in both the positively and negatively phrased items, indicating a stability characteristic of attitudes in Hennessy's terms.

The young adults in the 1978 study expressed near unanimity in regard to public regulation of the manufacture and sale of chemicals (see table 67). Approximately 90 percent of all educational status groups and both attentives and nonattentives rejected the idea that chemical manufacturers should be free of "governmental interference."

The 1978 data also indicated that a majority of young adults favored the space program, but that the proportion rejecting a negative statement about that program was significantly higher among students attentive to organized science than others. Within both the attentive and nonattentive groups, the proportion of college students favoring the space program was significantly higher than the proportion of college-bound high school students (see table 67). Generalizing from cross-sectional data, it appears that one effect of the college experience was to increase the appreciation of the space program among young adults, and the margin of increase appeared to be significantly larger for young adults who were already attentive to organized science than others.

Finally, it appeared that relatively few young adults believed that the United States trailed the Soviet Union and other nations in scientific achievement. A significantly higher proportion of college-bound high school students agreed with the view that the United States had lost its scientific leadership position in the world than college students, regardless of attentiveness. There were no significant differences between attentives and nonattentives in regard to this view. Although the United States was not thought to be trailing the Soviet Union in scientific achievement, a majority of the young adults in the 1978 study thought that it was important for the U.S. to continue to develop new weapons at least as fast as the Russians. The proportion of college students holding this view was significantly lower than the proportion of college-bound high school students, but those college students who were attentive to organized science were more likely to hold this view than nonattentive college students.

It is also possible to examine the spending preferences of the young adults in the 1978 study. The NPAS questionnaire included a list of science and technology spending areas and each respondent was told that these were areas for which "federal taxes are sometimes used." The respondent was asked to select the three areas that he or she most favored for support with tax funds and the three areas that were least favored.[18]

An analysis of spending preferences revealed a high degree of consensus among these young adults, with only minor deviations among educational status and attentiveness groups. Energy research was the most-favored object of science spending among all attentives and among nonattentive college students (see table 68). Science spending to reduce and control pollution was the second most-favored area among all attentives and among non-college-bound high school

TABLE 68
YOUNG ADULT PREFERENCES FOR SCIENCE AND TECHNOLOGY SPENDING: 1978

	Percentage Naming Area as One of Three "Most-Liked" Areas of Science Spending.					
	Attentive			Nonattentive		
	HSNC	HSCB	COLL	HSNC	HSCB	COLL
Energy	77%	68%	74%	32%	44%	59%
Pollution	47	38	51	34	33	46
Education	32	35	38	28	42	47
Health	22	23	37	27	27	37
Crime	39	35	27	59	55	35
Space	10	21	18	8	12	9
Birth control	6	14	14	18	14	17
Transportation	7	5	12	9	5	10
Weapons	25	28	11	16	19	9
Knowledge	3	9	10	9	7	10
Drug abuse	8	9	3	20	15	7
Weather	6	5	2	4	4	1
N =	28	287	662	1190	910	760

HSNC = high school student, no college plans
HSCB = high school student, college-bound
COLL = college student

[18] In the context of Hennessy's differentiation between opinions and attitudes, it is unlikely that these spending preferences represent strongly rooted attitudes for any except college attentives. The data are important in that they indicate the facets of science and technology that are most appealing to this generation.

students. Organized science spending to improve education and health care and to reduce crime were most favored by significant portions of young adults in all educational status and attentiveness groups.

The discovery of new basic knowledge about man and nature ranked relatively low in comparison to the more applied objectives of energy, pollution control, education, health care, and crime. Only ten percent of college students, regardless of attentiveness to organized science, mentioned this basic knowledge area as one of their three most-favored spending objectives, and it was mentioned by even fewer high school students.

In summary, these young adult spending preferences among science and technology objectives closely parallel adult objectives on comparable items. This result suggests strongly that the applied emphasis previously noted in the analysis of adult attitudes toward science policy is likely to continue. Even among college attentives, who presumably would have had the greatest exposure to the arguments for basic scientific research, there was little evidence of a recognition of the relationship between basic and applied research. The young adults generally held science and technology in high regard, stressing the benefits over the potential harms. There was support for increased federal spending for scientific research, but the spending preferences tended toward the solution of short-term and applied problems. The level of concern about the safety of nuclear power plants was somewhat higher than the comparable adult figures, but a significantly lower proportion of young adults who were attentive to organized science expressed concern about nuclear power than did young adults who were not attentive. There was optimism about a long-term solution to the energy problem from science and technology and a willingness to support weapons development needed to "keep up" with the Soviet Union. Contrary to leadership concerns, there was no evidence of a substantial antiscience group among these young adults and no support for the idea that the next generation is turning away from science and technology.

FUTURE POLITICAL PARTICIPATION

In the discussion of adult political behavior in earlier chapters, it was suggested that there has been increasing competition for each individual's time and that political concerns on the whole had not fared well in this marketplace. It is appropriate to look briefly at the dispositions of the young adults in the 1978 study toward political participation generally and possible participation in science policy issues more specifically.

In general, the young adult attitudes reported in the 1978 NPAS are fully consistent with the general models of interest, participation, and specialization presented in previous chapters. Young adults who were attentive to organized science were significantly more likely to report a high level of interest in news, current events, and politics than were comparable students who were not attentive to science matters (see table 69). There was broad agreement among the 1978 respondents, regardless of attentiveness, that most citizens are not well informed enough to make useful input into science policy decisions but that the interested and informed citizen could have some influence on the formulation of science policy if he or she were willing to make an effort. Among high school students, nonattentive students were more likely to agree that science policy should be left to scientists and other science policy experts than were attentive students, but this difference did not occur among college students. Generalizing from cross-sectional data, it would appear that one of the effects of college was

TABLE 69
YOUNG ADULT ATTITUDES TOWARD POLITICAL PARTICIPATION: 1978

	Attentive			Nonattentive		
Percentage who . . .	HSNC	HSCB	COLL	HSNC	HSCB	COLL
Are interested in news and current events	27%	50%	54%	14%	26%	31%
Scored high on Index of Political Interest	40	70	66	51	63	58
Agree that most citizens are not well informed enough to make useful input to policy decisions concerning science and technology	78	80	76	71	79	75
Believe that the interested and informed citizen can often have some influence on science policy decisions if he is willing to make the effort	66	63	69	61	64	62
Agree that science policy decisions should be left to scientists and other science policy experts	38	35	34	45	46	34
N =	28	187	662	1190	910	760

HSNC = high school student, no college plans
HSCB = high school student, college-bound
COLL = college student

to decrease the reliance on experts by those persons who were not personally attentive to science matters.

In broad strokes, the young adult generation measured by the 1978 NPAS were slightly more likely to be attentive to organized science than previous generations and tended to hold general dispositions and specific policy attitudes similar to previous generations. This new generation does not appear to be markedly more politically active than the older age groups discussed previously. There is no evidence of any generational disturbance in the operation of the political-specialization processes discussed above as they pertain to the formulation of science policy in the United States.

Chapter 10
The Formulation of Science Policy in a Democratic Society

There is inherent tension between the democratic values of our political system and the stratified structure of public participation in the formulation of science policy. Despite the analysis of the preceding chapters concerning the necessity, even the imperative, for a high degree of political specialization in modern societies like the United States, there is an uneasiness about a system in which two or three thousand policy leaders and a fifth of the public make policy decisions that have broad societal import. Recognizing this tension, this chapter will explore the question of whether the stratified structure of public participation is inherently elitist and whether there is an optimal size for the attentive public for science policy. The chapter will conclude with an examination of the utilization and effectiveness of the attentive public in the policy-formulation process.

IS AN ATTENTIVE PUBLIC INHERENTLY ELITIST?

The issue of whether the attentive public concept is inherently elitist concerns not only the size of attentive publics but, more importantly, their composition. Although it would be possible to argue on philosophical grounds that all binding political decisions ought to be made by at least a majority of the citizenry, the pressures discussed above in regard to the political-specialization process indicate that this alternative is not feasible. If some degree of political specialization is accepted as a necessary condition of modern life, then the size of the attentive public becomes relatively less important, provided that the proportion of persons attentive to any given issue is large enough to allow for some diversity of representation. In quantitative terms, it would not be reasonable to expect a majority of the American public to qualify as attentive to science policy, but, at the other extreme, an attentive public of only 1 or 2 percent would be too small to capture the diversity of public attitudes concerning an important issue domain like science policy.

The composition of an attentive public is central to the issue of elitism. If an attentive public were composed of only an exclusive segment of society or closed to large portions of the public, then that attentive public well might be considered elitist and would be offensive to the concept of democratic government. For example, if the attentive public for foreign policy were composed exclusively of persons with graduate degrees in political science or who were occupationally involved in foreign trade, it would take on an exclusive and self-serving character that would be largely inconsistent with the idea of a democratic political system. Since, in the first instance, attentiveness as defined in this analysis is driven by issue interest alone and since issue interest is not limited, or limitable, by degrees held or occupations practiced, the exclusivity or elitism implied in the preceding example would be most unlikely to occur in the American political system.

Looking at the attentive public for science policy, the analysis of its composition in chapter 4 indicated that it reflected age, gender, and educational diversity. For example, although the attentive public for science policy was somewhat better educated than other segments of the population, a majority of the attentive public did not have a college degree in either 1979 or 1981. Similarly, the proportion of males who were attentive to science policy was significantly higher than the proportion of females, but women made up about 40 percent of the attentive public for science policy in 1981. From these data, it is clear that the attentive public for science policy is not a perfect cross-section of the total public, but neither is it a major distortion of the composition of the American public. It included some persons from all educational and occupational groups and reflected a considerable range of age and regional variation. It would be very difficult to look at the demographic characteristics of the approximately 34 million Americans who are included in the attentive public for science policy and label that group elitist. Similar analyses of the demographic composition of the attentive publics for foreign policy, energy policy, local government, and local school politics have demonstrated a similar demographic diversity.

In summary, setting aside the issue of size *per se*, an examination of the demographic composition of the attentive public for science policy would not support the conclusion that this is an elitist group. More importantly, a consideration of the central role of personal issue interest in the definition of attentiveness and the unlimitable and uncontrollable nature of personal interest suggest that the concept is inherently pluralistic.

IS THERE AN OPTIMAL SIZE FOR AN ATTENTIVE PUBLIC?

In the preceding section, it was argued that the nature of the political-specialization process negated the likelihood of majority participation on issues like science policy and that the size of an attentive public was not evidence of elitism.

At the same time, it was suggested that there was some level below which the size of the attentive public would be unacceptable. This section will rejoin that issue and inquire as to whether it is possible to define an optimal size for an attentive public in general and for the attentive public for science policy in particular.

It is necessary to begin with an examination of the pattern of growth in the attentive public in recent years. The only available source of data concerning the size and composition of the attentive public in previous years is a 1957 survey of the interest of the public in science news and the public's knowledge about scientific concepts and issues.[19] The interviewing for the survey was completed only a matter of weeks before the launching of Sputnik I, making it the only available measure of public attitudes toward, and knowledge of, science in the pre-Sputnik period. Although the items were not identical to those used in the 1979 and 1981 studies, there was sufficient commonality to allow the construction of a comparable measure of attentiveness to science policy.

A comparison of the level of attentiveness to science policy in 1957 with the later measures indicated that the intervening 22 years had witnessed a modest growth in attentiveness and a substantial growth in interest in science-related matters (see table 70). An analysis of these changes within demographic classifications indicated that most of the growth in attentiveness to science policy occurred among persons who were under 35, males, and college graduates. In

TABLE 70
ATTENTIVENESS TO, AND INTEREST IN, SCIENCE POLICY: 1957–1981

	Attentive Public			Interested Public			N		
	1957	1979	1981	1957	1979	1981	1957	1979	1981
Total	14%	19%	20%	8%	20%	19%	1919	1472	2942
Age									
21–34	15	27	26	8	18	16	584	507	1068
35–54	15	16	20	9	21	19	853	492	1033
55 and over	9	13	13	5	21	22	470	473	843
Gender									
Female	11	12	15	8	21	20	1091	776	1568
Male	17	26	26	7	18	19	828	696	1375
Education									
Less than HS	5	5	7	6	17	17	1091	437	552
HS graduate	22	18	16	11	22	23	678	797	1746
Coll grad	40	46	43	7	17	12	144	238	645

[19] Sponsored by the National Association of Science Writers, the 1957 survey was designed to measure the public's interest in, and knowledge about, science and technology. The study was conducted by the Survey Research Center at the University of Michigan and obtained interviews with 1919 adults. For a technical description of the sample design, instrument, and data-collection procedures, see the Appendix and Davis (1958).

contrast, the growth of the interested public for science policy cut across all age, gender, and educational groupings. These data point to several important conclusions.

First, interest in science-related public policy matters appears to be broadly based and to be increasing among all sectors of the population. Since a high level of interest is a prerequisite for attentiveness, this pattern of interest indicates that science issues compete relatively well in the political-specialization context. In short, the major barrier to broader public participation in science policy does not appear to be the level of interest in science-related matters.

Second, the differential patterns of growth in attentiveness to science policy suggests that there are parallel differences in the distribution of knowledge about scientific concepts and issues. The largest and most obvious difference concerns the level of formal education completed. The level of attentiveness was highest among college graduates in all three years, and the marginal increase from 1957 was also the largest among this group. It should not be surprising that college graduates are better informed about scientific concepts and issues than other segments of the population. In contrast, the slight decline in the proportion of high school graduates who were attentive to science policy and the sharp increase in the proportion of high school graduates who were interested in science-related matters suggests that high school graduates were more likely to be deficient in knowledge about scientific concepts and issues in 1979 and 1981 than they were in 1957. Although aggregate in nature, this result raises questions about the quality of the high school science experience over the last two decades.

These education-related differences account for a large portion of the differences observed by age group, but they do not account for all of the gender-related differences in attentiveness. There was no significant gender-related difference in interest in science policy matters. Since the level of knowledge of scientific concepts and issues is the only difference between the attentive and interested publics, it would appear that the lower proportion of women who were attentive in all three years and the lower rate of increase in attentiveness must be associated with the acquisition of scientific information.

To examine the relative impact of gender in the context of educational and age differences, a multivariate stepwise logit analysis was performed, following the procedures described in previous sections. Attentiveness was treated as the dependent variable and dichotomized into attentives and nonattentives. Age was treated as a trichotomous variable and gender as a dichotomous variable. Education and exposure to a post-high school science course were combined into a trichotomous variable: less than college with no post-high school science courses, less than college with at least one post-high school science course, and college graduates.

The results of the stepwise logit analysis indicated that the new educational typology accounted for 80 percent of the total mutual dependence in 1957 and 70 percent in 1979 (see table 71). With educational level and course exposure

TABLE 71
A LOGIT MODEL TO PREDICT ATTENTIVENESS TO SCIENCE POLICY: 1957, 1979

Models		CMPD	
		1957	1979
H1	TAG,YT.	.804	.703
	Difference due to YT	.804	.703
H2	TAG,YT,YA.	.860	.747
	Difference due to YA	.056	.044
H3	TAG,YT,YA,YG.	.946	.890
	Difference due to YG	.086	.143

Y = attentiveness to science policy
T = educational typology
A = age
G = gender
CMPD = coefficient of multiple-partial determination

held constant, age differences accounted for about 5 percent of the total mutual dependence. With the educational typology and age held constant, the residual association between gender and attentiveness accounted for 9 percent of the total mutual dependence in 1957 and about 14 percent in 1979.

This result indicates that the gender difference in attentiveness to science policy has grown both numerically and in relative importance during the last two decades. Since there was no difference in the proportion of males and females in the interested public for science policy, the residual gender difference must be attributed to differences in knowledge about scientific constructs and issues. In a previous analysis of the 1978 survey of young adults, Miller, Suchner, and Voelker (1980) concluded that gender differences in attentiveness were attributable to early sex-role socialization and to a significantly lower level of peer interest in politics among young women. Whatever its roots, this gender difference should be addressed by science educators.

Projecting these patterns of growth into the future, it would appear that the size of the attentive public for science policy will continue to increase gradually, reflecting primarily the proportion of young adults attending college or participating in other post-high school science course experiences. If the trend of the last two decades were to carry forward, the attentive public for science policy would include about a quarter of the adult population by 1990 and perhaps 30 percent by the turn of the century. The total size of the attentive public would expand toward 40 million adults. This growth pattern would continue the current diversity of the attentive public but would gradually increase the proportion of college graduates and advanced degree holders. It would not, however, become grossly unrepresentative of the American people.

Having examined the present trends in regard to the size and composition of the attentive public, it is appropriate to return to the original question of whether there is an optimal size for an attentive public generally and for the attentive public for science policy in particular. At a general level, it would appear that an attentive public should be sufficiently large and broadly distributed to assure that, when mobilized, it could influence enough legislators to effectively veto policies detrimental to its interest. For some issue domains, like agriculture, the geographic distribution of that attentive public and the configuration of congressional committee memberships relevant to those interests combine to require a relatively small attentive group. Other issues, like foreign policy or science policy, lack the regional characteristics of agriculture and must be able to seek to influence legislators from all regions of the country. For "national" issues, a larger attentive public is required to effectively protect its interests. While there is no algorithm to estimate the minimal size of an attentive public, the considerations discussed above can be employed to assess the attentive public for any particular issue.

Applying these considerations to the attentive public for science policy, it would appear that this attentive public already exceeds the minimum size needed to protect its interest. Using the previous estimate of about 34 million adults, a mobilization rate of 10 percent would yield 3.4 million contacts with decision makers and a mobilization rate of 20 percent would produce nearly seven million citizen contacts. For most purposes, this would be a sufficient volume of activity to assure the protection of the status quo. The continued growth of the attentive public for science policy along the patterns suggested above would assure a continuing ability to defend the interests of the scientific community.

It is reasonable to inquire what size attentive public would be needed to modify the status quo in favor of organized science. The question and the issues it raises are much more complex. At the present time, the major issue on the science policy agenda is a larger share of federal resources for scientific research and education generally and for basic scientific research in particular. In the current federal budget-making process, the acquisition of a significantly larger share of funding would require either a higher federal deficit, higher federal taxes, or a share of resources previously allocated to another purpose or program. All three of these options would activate other attentive publics that would wish to either preserve their share of federal funding or avoid new taxation. In the context of high unemployment and other competing social needs, it is unlikely that organized science could generate an attentive public large enough to produce sufficient direct pressure to force a reallocation of federal resources in favor of science. In the context of the pressure controversies over federal spending and the allocation of the federal budget, it is likely that a united science policy leadership group would be more effective in persuading both executive and legislative decision makers of the need for additional resources.

Apart from current federal budget policy, the need for an attentive public for science policy large enough to directly influence the establishment of new policies is not clear. On most matters that are technical in character, science policy leaders are more effective in presenting and negotiating for specific policy outcomes than the broader and more diverse attentive public. In very general terms, for policies other than the acquisition of resources, the primary use of the attentive public for science policy is defensive, that is, to protect against the erosion or reversal of the postwar relationship between the government and organized science.

In summary, there is no single optimal size for an attentive public. A minimal size for an attentive public can be defined conceptually, depending on the regional or national character of the issue and the nature of the policy demands. For national issues like science policy, it is necessary to have a sufficiently large attentive public to be able to generate contacts with a broad spectrum of legislators, especially for the purpose of protecting current policies. It would appear that the attentive public for science policy is, and will remain, large enough for that purpose.

THE UTILIZATION AND EFFECTIVENESS OF THE ATTENTIVE PUBLIC

The most important questions concerning the attentive public for science policy are the frequency of its utilization and its effectiveness in the policy-formulation process. This section will explore the frequency of leadership efforts to mobilize the attentive public for science policy and then seek to assess the efficacy of the contacting and other interventions by the mobilized portion of the attentive public.

Looking first at utilization, science policy leaders have rarely attempted to mobilize the attentive public for science policy. The problem is twofold. First, the nongovernmental leadership of science policy has been successful in obtaining most of the policy objectives of organized science in the postwar years through direct negotiation with decision makers. In the conflicts over the establishment of the Atomic Energy Commission and the National Science Foundation, science policy leaders did appeal for broader public pressure but mainly through disciplinary and professional associations. An attentive public of the current size and scope was not available in the late 1940s and the influence of scientific leaders fresh from their wartime achievements was sufficient for most purposes. The acquisition of additional resources during the 1950s and in the first decade after Sputnik did not require additional public pressure.

Despite the slippage in real purchasing power for scientific research during the 1970s, science policy leaders did not attempt to mobilize broader public support for the acquisition of additional resources. Current dollar increases during this period were relatively large, and the full impact of inflation did not become apparent until later in the decade. The first overt attempt to mobilize broader public support for scientific research and for science education came in response to the substantial domestic-spending reductions proposed by the incoming Reagan administration.

The second part of the problem concerns the attitudes of at least a portion of the science policy leadership concerning the scope of public participation in the formulation of science policy. Even though most science policy leaders agreed with the statement that the interested and informed citizen could have some influence on science policy, only half of the leadership group reported any effort to mobilize nonscientists who were interested in, and informed about, science policy: the attentive public. An examination of the mobilization activities of the major disciplinary societies and professional associations indicates that most of these organized efforts focus on their own membership, with little or no overt effort to identify and mobilize the 30 million attentives in the broader population. The utilization of the attentive public for science policy must await, in part, the development of greater awareness of the existence of these citizens by science policy leaders and a willingness to open the policy-formulation process to nonscientists.

Given the infrequent efforts of science policy leaders to mobilize the attentive public, it is difficult to assess its efficacy. The most recent effort to mobilize attentive participation in the policy process was the activities of the major disciplinary societies and professional associations to protect science funding from prospective budget reductions in domestic programs proposed by the incoming Reagan administration. A major portion of this effort was designed to persuade the members of scientific groups to contact congressional decision makers to emphasize the importance of funding for scientific research. There were only a few efforts to reach out beyond the scientific community itself, and the results of the mobilization effort were correspondingly modest. The data from the 1981 study indicated that only 5 percent of the attentive public reported contacting a public official on a science-related matter in the previous year, and a detailed analysis of the subject matter of those contacts revealed that many of the contacts concerned nonresource issues or were contacts with state or local officials on other matters. A realistic estimate of the mobilization of the attentive public concerning science funding would be 1 or 2 percent. This is a disappointing level of mobilization by any yardstick. The fact that scientific research suffered relatively smaller budgetary reductions than other domestic activities in the first Reagan budget should be viewed as the result of effective lobbying by science policy leaders directly rather than as a measure of the utilization or effectiveness of the attentive public.

A parallel mobilization effort was made to prevent the elimination of federal science education programs by the Reagan administration. In this case, the mobilization effort involved both science policy organizations and educational associations with concerns broader than science. While solid data on the level of mobilization are not available in regard to the science education controversy, it would appear that pressure from the mobilized public helped persuade a sufficient number of House members of the value of the program to protect it within the committee structure. The recommendations from the Committee on Science and Technology and the Appropriations Committee were both favorable to science education. In this case, the Reagan strategy of placing a substitute budget bill before the House of Representatives after the conclusion of the normal committee review process successfully negated the efforts of the mobilized segment of the attentive public for science policy to preserve a federal program in science education. Using the parallel mobilization effort to preserve scientific research funding as a yardstick, it is likely that the rate of mobilization on the science education issue was at the 2 to 3 percent level.

From this brief review, it would appear that there have been relatively few efforts by science policy leaders to mobilize the attentive public in the postwar years and that those few attempts have not been very successful. The high rate of success of science policy leaders in obtaining their major objectives without broader public pressure has resulted in a leadership group largely unskilled in mobilization and an attentive public unaccustomed to being mobilized. If the success of science policy leaders in direct negotiation could be assured for the future, there would be no occasion for concern. That assumption, however, is risky.

The reform of the federal budget process in the latter part of the 1970s effectively changed the resource allocation process into a zero-sum situation. Each sector of society that seeks a share of federal resources must compete against other claimants in *both* the substantive appropriations subcommittee and in the general budget committee. During periods of overall budgetary growth, the competition for resources is vigorous but manageable. During periods of stable or declining budget ceilings, the competition for resources becomes much more intense and hostile. For a variety of political and economic reasons, it is likely that the aggregate level of federal resources, in real dollar terms, will be relatively stable over the next decade and perhaps beyond. It is very likely that the competition for those resources will intensify.

The impact of political specialization on American politics has been substantial and, undoubtedly, will be even greater in the decade ahead. In this political milieu, science policy leaders will need to mobilize their attentive public more frequently and more effectively. The cultivation and mobilization of an attentive public is a continuous long-term process. If the interests of organized science are to be advanced, it is essential that both science policy leaders and the attentive public for science policy improve their understanding of the policy-formulation process and learn to work cooperatively in that process.

Appendix
The Data Base

The analysis reported in this book is based primarily on four national surveys conducted since 1978. A brief description of those studies will be helpful in understanding the data analyses and references throughout the book.

Most of the analysis relies on two national probability surveys of the adult population of the United States that were conducted in 1979 and 1981 under a contract and a grant, respectively, from the Science Indicators Unit of the National Science Foundation for use in the biennial National Science Board report *Science Indicators*.

In 1972, 1974, and 1976, the NSF had sponsored a set of science and technology attitude questions as a part of an omnibus survey and reported the results in the first three issues of *Science Indicators* (NSB, 1973, 1975, 1977). There was general recognition by both the NSF staff and the social science community that these surveys were too limited in scope and without a solid theoretical basis. In 1978, the NSF issued a "request for proposal" (RFP) asking the social science and survey research communities to propose a new approach to these surveys. Miller and Prewitt were selected by the NSF to design the survey instrument for the 1979 survey and to analyze the resulting data. A separate RFP was issued for the field work and the Institute for Survey Research (ISR) at Temple University was selected for that phase of the work, utilizing the instrument designed by Miller and Prewitt in phase one.

The 1979 survey was a multistage cluster sample that produced a total of 1635 interviews. The ISR sample included approximately one hundred primary sampling units and a completion rate of approximately 75 percent was obtained. The interviews were conducted by interviewers visiting the respondent's home and the interviews averaged 55 minutes in length. For a technical description of the sampling design, the response rates, weighting procedures, and instrument, see Miller, Prewitt, and Pearson (1980).

While the 1979 survey provided important baseline data concerning the structure of public participation in science policy, there were numerous additional questions that could not be included in the original study. In 1980, Miller and Prewitt proposed a second survey to the NSF that would allow a more in-depth analysis of both the substantive policy attitudes of various segments of the public and of their actual policy-participation behaviors. The NSF funded the proposed survey and in 1981 the Public Opinion Laboratory at Northern Illinois University conducted a national survey of adult attitudes toward science and technology. The 1981 survey used a multistage cluster sample comparable to the 1979 study, but the 1981 sample utilized a total of 150 primary sampling units. The 1981 survey employed a screen technique that allowed the identification of respondents with a high level of interest in, and knowledge about, science and technology during the first 10 minutes of the interview. Using "computer-assisted-telephone-interviewing" (CATI) technology, it was possible to continue with a more sophisticated set of items for those respondents who demonstrated an adequate level of interest and knowledge and to conclude the interview for the other respondents. Of a total of 3195 interviews, approximately 40 percent were selected for, and completed, the in-depth interview. An overall completion rate of 80 percent was obtained. For a technical description of the sampling design, response rates, weighting procedures, and instrument, see Miller and Prewitt (1982b).

As a part of their 1980 proposal to the NSF, Miller and Prewitt outlined a separate survey of the policy attitudes and political behavior of a sample of the leaders of American science and technology. The NSF approved the request.

Since there had been no previous attempts to define or survey this group, there were no existing listings of persons or sets of operational definitions. Following the general structure described by Price (1965) and others, the universe of persons considered to be science and technology policy leaders for the 1981 survey included all persons who either:

1. were officers of national professional or scientific societies or associations;
2. had testified before a committee of the Congress on a scientific or technological issue within the previous two years;
3. had served on a major executive branch science or technology policy advisory committee within the previous two years;
4. had published a book or refereed journal article on a science or technology policy matter within the previous two years;
5. were members of the National Academy of Sciences, the National Academy of Engineering, or the Institute of Medical Sciences;
6. had won a Nobel Prize in a scientific field or a Field Medal (mathematics);
7. had served as officers or on the board of directors of one of the 20 top science or engineering corporations on the Fortune 500 list; or
8. were full-time science journalists for a national distribution broadcast or print organization.

An individual's name was entered into the universe list once for each time that he or she qualified; thus, some individuals were entered into the universe list several times while others were entered only once. The application of these criteria produced a universe list of approximately 6500 entries.

From this universe list, a random sample of 360 names was selected. Twenty-two of the original names were dropped from the study for reasons ranging from absence from the country during the period of the study to the death of a respondent. Of the adjusted universe list of 338 respondents, interviews were completed with 281 individuals. For any individual, the probability of selection was proportionate to the frequency that his or her name was in the file. This "probability-proportionate-to-frequency" (PPF) principle provided a simple weighting procedure to take into account differences in the level of leadership activity of various individuals.

The interviews were conducted by telephone from the Northern Illinois University Public Opinion Laboratory and required about 30 minutes.

The fourth survey utilized in the analysis is a 1978 national survey of high school and college students in the United States. Funded by a grant from the science education program of the National Science Foundation, the 1978 National Public Affairs Study was designed to test a broad set of hypotheses concerning attitude formation about political issues. Designed by Miller, Suchner, and Voelker, the study was based on the on-site administration of questionnaires to approximately three thousand high school students and the collection of mailed questionnaires from approximately one thousand college students in four-year institutions. For a technical description of the sample design, response rates, weighting procedures, and instrument, see Miller, Suchner, and Voelker (1980).

The combination of these four surveys represents the broadest and most current data base available concerning the attitudes of the public, adult and young adult, and the nature and frequency of public participation in the formulation of policies relevant to science and technology.

REFERENCES

Allport, G. 1954. *The nature of prejudice*. New York: Anchor Doubleday.

Almond, G. 1950. *The American people and foreign policy*. New York: Harcourt, Brace, & Company.

Almond, G., and S. Verba. 1963. *The civic culture*. Boston: Little, Brown, & Company.

Astin, A. 1977. *Four critical years: Effects of college on beliefs, attitudes, and knowledge*. San Francisco: Jossey-Bass.

Bronfenbrenner, U. 1970. *Two worlds of childhood*. New York: Russell Sage Foundation.

Brooks, H. 1978. The problem of research priorities. *Daedalus* 107:2:171-190.

Bush, V. 1945. *Science: The endless frontier*. Washington, D.C.: U.S. Government Printing Office.

Campbell, A., P. Converse, W. Miller, and D. Stokes. 1960. *The American voter*. New York: Wiley.

Cohen, B. C. 1959. *The influence of nongovernmental groups in foreign policy-making*. Boston: World Peace Foundation.

————————. 1973. *The public's impact on foreign policy*. Boston: Little, Brown, & Company.

Conant, J. B. 1952. *Modern science and modern man*. Garden City, NJ: Masterworks Program.

Converse, P. 1964. The nature of belief systems in mass publics. In *Ideology and discontent,* ed. D. E. Apter. New York: Free Press.

————————. 1970. Attitudes and non-attitudes: Continuation of a dialogue. In *The qualitative analysis of social problems,* ed. E. R. Tufte. Reading, MA: Addison-Wesley.

Davis, R. C. 1958. *The public impact of science in the mass media*. Survey Research Center, Monograph no. 25, University of Michigan.

Dawson, R., K. Prewitt, and K. Dawson. 1978. *Political socialization*. Boston: Little, Brown, & Company.

Easton, D., and J. Dennis. 1969. *Children in the political system*. New York: McGraw-Hill.

Etzioni, A., and C. Nunn. 1974. The public appreciation of science in contemporary America. *Daedalus* 3:191-205.

Fienberg, S. E. 1980. *The analysis of cross-classified categorical data*. 2d ed. Cambridge, MA: Massachusetts Institute of Technology Press.

Goodman, L. 1972a. A general model for the analysis of surveys. *American Journal of Sociology* 77:1035-1086.

————————. 1972b. A modified multiple regression approach to the analysis of dichotomous variables. *American Sociological Review* 37:28-46.

————————. 1978. *Analyzing qualitative/categorical data: Log-linear models and latent structure analysis*. Cambridge, MA: Abt Books.

Greenberg, D. 1967. *The politics of pure science*. New York: Plume.

Hennessy, B. C. 1966. Public opinion and opinion change. In *Political science annual: An international review,* Vol. 1. ed. J. A. Robinson. New York: Bobbs-Merrill.

————————. 1972. A headnote on the existence and study of political attitudes. In *Political attitudes and public opinion,* ed. D. D. Nimmo and C. M. Bonjean. New York: McKay.

Hero, A. O. 1959. *Americans in world affairs*. Boston: World Peace Foundation.

————————. 1960. *Voluntary organizations in world affairs*. Boston: World Peace Foundation.

Hess, R. D., and J. V. Torney. 1965. *The development of political attitudes in children*. Chicago: Aldine.

Hewlett, R. G., and O. E. Anderson, Jr. 1962. *The new world, 1939/1946: A history of the United States Atomic Energy Commission*. University Park, PA: The Pennsylvania State University Press.

Hyman, H. H. 1959. *Political socialization*. New York: Free Press.

Hyman, H. H., C. R. Wright, and J. S. Reed. 1975. *The enduring effects of education*. Chicago: The University of Chicago Press.

Jennings, M. K., and R. G. Niemi. 1974. *The political character of adolescence*. Princeton, NJ: Princeton University Press.

——————. 1981. *Generations and politics*. Princeton, NJ: Princeton University Press.

Keniston, K., and the Carnegie Council on Children. 1977. *All our children: The American family under pressure*. New York: Harcourt, Brace, & Jovanovich.

Knoke, D., and P. J. Burke. 1980. *Log-linear models*. Beverly Hills, CA: Sage Publications.

Milbrath, L. 1965. *Political participation: How and why people get involved in politics*. Chicago: Rand McNally.

Miller, J. D. 1978. A stratified model of attitudes toward the politics of community planning. In *The small city and regional community,* Vol. 1. ed. R. Wolensky and E. J. Miller. Stevens Point, WI: Foundation Press, Inc.

——————. 1980. Political and issue specialization: A behavioral imperative. Paper presented at the meeting of the American Political Science Association, Washington, D. C.

——————. 1981. The politics of educational change. In *The NCTM resource book on change,* ed. J. Price. Los Angeles: National Council of Teachers of Mathematics.

——————. 1982a. *The information needs of the public concerning space exploration*. A special report prepared for the National Aeronautics and Space Administration.

——————. 1982b. *A national survey of public attitudes toward science and technology*. A report to the National Science Foundation under NSF grant 8105662. DeKalb, IL: Public Opinion Laboratory.

——————. 1983a. A conceptual framework for understanding public attitudes toward conservation and energy issues. In *Energy and material resources: attitudes, values, and public policy,* ed. D. Conn. Boulder, CO: Westview Press.

——————. 1983b. Scientific literacy: a conceptual and empirical review. *Daedalus* 112:2:29-48.

Miller, J. D., and K. Prewitt. 1979. *The measurement of the attitudes of the U.S. public toward organized science*. Report submitted to the National Science Foundation in accordance with contract SRS78-16839. Chicago: National Opinion Research Center.

——————. 1982. *A national survey of the nongovernmental leadership of American science and technology*. A report to the National Science Foundation under NSF grant 8105662. DeKalb, IL: Public Opinion Laboratory.

Miller, J. D., K. Prewitt, and R. Pearson. 1980. *The attitudes of the U.S. public toward science and technology*. Report submitted to the National Science Foundation. Chicago: National Opinion Research Center.

Miller, J. D., R. W. Suchner, and A. M. Voelker. 1980. *Citizenship in an age of science*. New York: Pergamon.

Mueller, J. E. 1973. *War, presidents, and public opinion*. New York: Wiley.

National Science Board. 1973. *Science indicators—1972*. Washington, D.C.: U.S. Government Printing Office.

——————. 1975. *Science indicators—1974*. Washington, D.C.: U.S. Government Printing Office.

——————. 1977. *Science indicators—1976*. Washington. D.C.: U.S. Government Printing Office.

——————. 1981. *Science indicators—1980*. Washington, D.C.: U.S. Government Printing Office.

Nelkin, D. 1978. Threats and promises: negotiating the control of research. *Daedalus* 107:2:191-210.

Newcomb, T. M. 1976. *Persistence and change: Bennington college and its students after twenty-five years*. New York: John Wiley and Sons.

Nie, N., and K. Anderson. 1974. Mass belief systems revisited: Political change and attitude structure. *Journal of Politics* 36:540-591.

Nie, N., S. Verba, and J. Petrocik. 1976. *The changing American voter*. Cambridge, MA: Harvard University Press.

Oppenheim, K. 1966. *Acceptance and distrust: Attitudes of American adults toward science*. Unpublished Master's Thesis, University of Chicago.

Penick, J. L., C. W. Purcell, M. B. Sherwood, and D. C. Swain. 1965. *The politics of American science: 1939 to the present*. Rev. ed. Cambridge, MA: Massachusetts Institute of Technology Press.

Price, D. K. 1954. *Government and science*. New York: New York University Press.

———————. 1965. *The scientific estate*. Cambridge, MA: Belknap Press of Harvard University Press.

RePass, D. E. 1971. Issue salience and party choice. *American Political Science Review* 65:389-400.

Rosenau, J. 1961. *Public opinion and foreign policy: An operational formulation*. New York: Random House.

———————. 1963. *National leadership and foreign policy: The mobilization of public support*. Princeton, NJ: Princeton University Press.

———————. 1974. *Citizenship between elections*. New York: Free Press.

Rosenberg, M. 1957. Misanthropy and political ideology. *American Sociological Review* 21:690-695.

Rummel, R. J. 1972. *The dimensions of nations*. Beverly Hills, CA: Sage Publications.

Shen, B. 1975. Scientific literacy and the public understanding of science. In *Communication of scientific information,* ed. S. Day. Basel: Karger.

Stewart, I. 1948. *Organizing scientific research for war*. Boston: Little, Brown, & Company.

Sullivan, J. L., J. E. Piereson, and G. E. Marcus. 1978. Ideological constraint in the mass public: A methodological critique and some new findings. *American Journal of Political Science* 22:233-249.

Taviss, I. 1972. A survey of popular attitudes toward technology. *Technology and Culture* 13:606-621.

Toffler, A. 1970. *Future shock*. New York: Random House.

U.S. Congress, Senate, Subcommittee on War Mobilization. 1945. *The government's wartime research and development, 1940-44 Part II—Findings and recommendations*. Report to 79th Congress, 1st Session, Washington, D.C.

Verba, S., and N. Nie. 1972. *Participation in America: Political democracy and social equality*. Ann Arbor: University of Michigan Press.

Withey, S. B. 1959. Public opinion about science and the scientists. *Public Opinion Quarterly* 23:382-388.

Index

Acquisition of scientific information, 111
Almond, Gabriel, xv, 23, 39, 58
Atomic Energy Commission, 12–3
Attentive public,
  basic concept, xv
  definition, 23, 29
  elitism issue, 125–6
  optimal size, 126–31
  relationship to democratic theory, xv
Attentive public for science policy,
  change since 1957, 127–8
  characteristics of, 44–8
  definition, 39–43
  emergence during young adulthood, 106–15
  general dispositions toward science, 49–54
  policy impact, xv
  role awareness, 57
  size, 43
  social and political behavior of, 58–9
Attentive public for food additive issue,
  characteristics of, 82–5
  conflict resolution, 90–3
  definition, 79–82
  personal participation, 93–5, 101–3
  policy attitudes, 89–90
  size, 82
Attentive public for nuclear power issue,
  characteristics of, 82–5
  conflict resolution, 90–3
  definition, 79–82
  personal participation, 93–6, 98–101
  policy attitudes, 87–9
  size, 82
  young adult attitudes, 119–20
Attentive public for space exploration,
  characteristics of, 82–5
  conflict resolution, 90–3
  definition, 79–82
  personal participation, 93–8
  policy attitudes, 86–7
  size, 82
  young adult attitudes, 119–20
Attentiveness to specific issues, 79–81

Bohr, Nils, 8
Bronfenbrenner, Urie, 26
Brooks, Harvey, 71
Bush, Vennevar, 9, 12, 14–6, 18, 60

Committee on Medical Research, 9
Compton, Karl T., 7
Conant, James, 6, 9
Converse, Phillip, 27–8
Creationism, 77–8

Davis, Robert, 3–22

Einstein, Albert, 8

Fermi, Enrico, 8
Fienberg, Stephen, 44

Goodman, Leo, 44
Governmental regulation of pharmaceutical products, 73
Greenberg, Dan, 8

Hennessy, Bernard, 48

Independence of scientific inquiry, 4

control of nuclear research, 11–14
governmental regulation, 71–3
pre-war issues, 5–7
public willingness to restrain, 74–7
ranking on leadership agenda, 55–6, 71
World War II mobilization, 7–11

Interest specialization,
definition, 26–7
emergence during young adulthood, 122
factors contributing to, 25

Interested public for science policy,
attitudes toward creationism, 77–8
attitudes toward funding for scientific research, 61–5
characteristics of, 44
definition, 43
general dispositions toward science, 49–54
restraint of scientific inquiry, 74–5
size, 43–4

Issue specialization,
definition, 28
emergence during young adulthood, 106–9
factors contributing to, 25

Issue voting, 29

Jewett, Frank, 9

Keniston, Kenneth, 26
Kilgore, Senator Harley, 10–11, 14

Logit analysis, 44, 53–54, 74–6, 83–4, 87, 97–8, 114–5, 117–8, 128–9

Milbrath, Lester, 26–7
Miller, Jon, 22, 42, 72, 73, 105, 134
Mobilization, 24, 131–3
Mobilized public, 23

National Defense Research Committee (NDRC), 9
National Science Foundation (NSF), 14–21
Nelkin, Dorothy, 71
Nie, Norman, 27
Nobel prizes, 6
Non-attentive public for science policy,
attitudes toward creationism, 77–8
attitudes toward funding for scientific research, 61–5
definition, 48
general dispositions toward science, 49–54
potential influence of, 48–9
restraint of scientific inquiry, 74–7

Office of Science and Technology (OST), 34
Office of Science and Technology Policy (OSTP), 34
Office of Scientific Research and Development (OSRD), 9
university research contract, 10
Office of Technology Assessment (OTA), 34
Organized science,
definition, 2, 105
development of attentiveness to, 106–15

Political socialization, 105
Political specialization,
Factors contributing to, 22, 25
typology, 28–30
Prewitt, Kenneth, xvii, 2, 23, 25, 75
Price, Don K., 135
Public understanding of science, 2, 55, 71–3, 109–11

Referenda, 24–5
Restraint of scientific inquiry, 74–77
Risk-benefit assessments of,
science by young adults, 116–8
scientific research, 52–54
Roosevelt, President Franklin D., 89
Rosenau, James, 23, 58

Science court, 1
Science education
attentive public attitudes toward, 65–6

international comparisons, 66
leadership attitudes toward, 65–6
ranking on leadership agenda, 55
science knowledge among young adults, 110–11
Science Indicators, 134
Science policy,
agenda, 55–7
formulation of, 24, 57–9
Science policy decision-makers, definition, 24, 33
Science policy leaders,
agenda-setting role, 55
definition, 23–4
efforts to influence policy, 38–40
profile, 35–8
Scientific community, 4
Suchner, Robert, 3, 105, 113, 129
Support for scientific research,
attentive public attitudes toward, 61–5
as objective of scientific community, 4
basic versus applied emphasis, 63–4
creation of National Science Foundation, 14–7
impact of inflation, 19–20
impact of Sputnik, 19–20
impact of World War II, 7–11
industrial research and development, 67–8
international competition, 68–70
Kilgore Report, 10–11
leadership attitudes toward, 60–5
National Defense Research Committee (NDRC), 9
Office of Scientific Research and Development (OSRD),9
pre-war experience, 5–7
ranking on leadership agenda, 55–6
young adult attitudes, 118–20

Toffler, Alvin, 26

Verba, Sidney, 27, 58
Voelker, Alan, 3, 105, 113, 129

About the Author

JON D. MILLER is Director of the Public Opinion Laboratory and Associate Professor of Political Science at Northern Illinois University. He earned his baccalaureate at Ohio University, his master's degree at the University of Chicago, and his doctorate at Northwestern University. He has held a Danforth Foundation Fellowship, a National Opinion Research Center Fellowship, and is a member of Phi Beta Kappa.

For the last decade, Professor Miller has studied the development and structure of public attitudes toward science and technology. He is the author of *Citizenship in an Age of Science* (Pergamon, 1980), which focused on the development of attitudes toward science and technology during the young adult years. He has published numerous articles on public participation in the formulation of science policy and on scientific literacy. Professor Miller is a member of the Committee on the Public Understanding of Science of the American Association for the Advancement of Science.